Antimicrobial Resistance - A One Health Perspective

Edited by Mihai Mareș, Swee Hua Erin Lim, Kok-Song Lai and Romeo-Teodor Cristina

Published in London, United Kingdom

IntechOpen

Supporting open minds since 2005

Antimicrobial Resistance - A One Health Perspective
http://dx.doi.org/10.5772/intechopen.87316
Edited by Mihai Mareș, Swee Hua Erin Lim, Kok-Song Lai and Romeo-Teodor Cristina

Contributors
Suraja Kumar Nayak, Swaraj Mohanty, Bighneswar Baliyarsingh, Nitya Meenakshi Raman, Murugesh Easwaran, Rashmi Kaul, Jyotsna Bharti, Khaled Fathy Abdel Motelb, Tanushri Kaul, Carol Lopez De Dicastillo, Matias Guerrero Correa, Fernanda B. Martínez, Camilo Zuñiga, Maria José Galotto, Rosalino Vázquez-López, Sandra Solano - Gálvez, Diego Abelardo Álvarez-Hernández, María Fernanda Valencia-Segrove, María José Ostos Prado, Ana Berenice López Boucieguez, Miliane Souza, Cláudio Rocha-De-Souza, Dayanne Melo, Cássia Motta, Ramon Pimenta, Irene Coelho, Shana Coelho, Chandrajit Lahiri, Shama Mujawar, Bahaa Abdella, Stelian Baraitareanu, Livia Vidu, Mohammad Mahmudul Hassan, Cristina Paiva De Sousa, Felipe De Paula Nogueira Cruz, Andréa Cristina Bogas

First published in London, United Kingdom, 2021 by IntechOpen
IntechOpen is the global imprint of INTECHOPEN LIMITED, registered in England and Wales, registration number: 11086078, 5 Princes Gate Court, London, SW7 2QJ, United Kingdom
Printed in Croatia

British Library Cataloguing-in-Publication Data
A catalogue record for this book is available from the British Library

Additional hard and PDF copies can be obtained from orders@intechopen.com

Antimicrobial Resistance - A One Health Perspective
Edited by Mihai Mareș, Swee Hua Erin Lim, Kok-Song Lai and Romeo-Teodor Cristina
p. cm.
Print ISBN 978-1-83962-432-2
Online ISBN 978-1-83962-433-9
eBook (PDF) ISBN 978-1-83962-434-6

For EU product safety concerns:
IN TECH d.o.o., Prolaz Marije Krucifikse Kozulić 3, 51000 Rijeka, Croatia, info@intechopen.com or visit our website at intechopen.com.

Meet the editors

Dr. Mihai Mareș received his Ph.D. degree in Microbiology at Gr. T. Popa University of Medicine and Pharmacy from Iași-Romania (2005) and had postgraduate training at University VII Denis-Diderot, Pasteur Institute, Pitié-Salpêtrière Hospital, École du Val-de-Grâce - Paris (France), Complutense University – Madrid (Spain), Instituto de Salud Global - Barcelona (Spain), Karolinska Institute – Stockholm (Sweden), and Danish Technical University - Lyngby (Denmark). His areas of interest are medical mycology, antimicrobial resistance, mycobacteria, food microbiology, biofilms, microbial induced infertility, and bio-medical applications of plasma discharges and cold plasma activated water. Currently, Dr. Mareș is a Professor of Microbiology and Head of the Antimicrobial Chemotherapy Laboratory at Ion Ionescu de la Brad University – Iași (Romania). Also, he is a member of the EUCAST Antifungal Susceptibility Testing Subcommittee and ESCMID Study Group for Veterinary Microbiology. He has served as a scientific consultant for several pharmaceutical companies during the past few years.

Dr. Erin Lim is presently working as an Assistant Professor in the Division of Health Sciences, Abu Dhabi Women's College, Higher Colleges of Technology in Abu Dhabi, United Arab Emirates and is affiliated as an Associate Professor to Perdana University-Royal College of Surgeons in Ireland, Selangor, Malaysia. She obtained her Ph.D. from Universiti Putra Malaysia in 2010 with a National Science Fellowship awarded from the Ministry of Science, Technology and Innovation Malaysia and has been actively involved in research ever since. Her main research interests include analysis of carriage and transmission of multidrug resistant bacteria in non-conventional settings, besides an interest in natural products for antimicrobial testing. She is heavily involved in the elucidation of mechanisms of reversal of resistance in bacteria in addition to investigating the immunological analyses of diseases, development of vaccination and treatment models in animals. She hopes her work will support the discovery of therapeutics in the clinical setting and assist in the combat against the burden of antibiotic resistance.

Dr. Lai Kok Song is an Assistant Professor in the Division of Health Sciences, Abu Dhabi Women's College, Higher Colleges of Technology in Abu Dhabi, United Arab Emirates. He obtained his Ph.D. in Biological Sciences from Nara Institute of Science and Technology, Japan in 2012. Prior to his academic appointment, Dr. Lai worked as a Senior Scientist at the Ministry of Science, Technology and Innovation, Malaysia. His current research areas include antimicrobial resistance and plant-pathogen interaction. His particular interest lies in the study of the antimicrobial mechanism via membrane disruption of essential oils against multi-drug resistance bacteria through various biochemical, molecular and proteomic approaches. Ultimately, he hopes to uncover and determine novel biomarkers related to antibiotic resistance that can be developed into new therapeutic strategies.

Currently, Dr. Romeo-Teodor Cristina is a Professor of Veterinary Pharmacology at Banat's University of Agricultural Sciences and Veterinary Medicine "King Michael I of Romania" of Timișoara-Romania. He obtained his Ph.D. degree in 1997 and had a postgraduate instruction at Liverpool University - School of Veterinary Medicine, UK. His areas of interest are antimicrobial resistance, veterinary drug agents, therapy, phytotherapy, and pharmacovigilance. He is the Editor-in-chief of the journal "Veterinary Drug" and a corresponding member of the Romanian Academy of Agricultural and Forestry Sciences, the European Association for Veterinary Pharmacology and Toxicology, and the European Federation for Pharmaceutical Sciences - Network on Veterinary Medicines. During the past few years, he has served as a technical consultant for several pharmaceutical companies and veterinary national authorities.

Contents

Preface

Antimicrobial resistance (AMR) is fast becoming a formidable challenge globally not only in the clinical settings, but also in the agricultural and community settings. Currently, we are running low on options with the depleting antibiotic pipeline and understanding the enemy, in this case, the resistant mechanisms in microorganisms with relation to host interaction, is perhaps the most straightforward step to take in a measured, calculated attempt to solve this problem.

This book explores molecular mechanisms with regard to AMR, control strategies in agriculture, and closing with One Health challenges. It is unquestionable that AMR, inevitably, affects all aspects of life and as long as this cycle remains unbroken and is continuously evolving and expanding, the ability to maintain our very own existence remains threatened. Indeed, the pre-antibiotic era may be nearer than we think.

Mihai Mares
Ion Ionescu de la Brad University of Agricultural Sciences
and Veterinary Medicine of Iași,
Romania

Swee Hua Erin Lim
Health Sciences Division, Abu Dhabi Women's College,
Higher Colleges of Technology,
Abu Dhabi, UAE

Kok Song Lai
Health Sciences Division, Abu Dhabi Women's College,
Higher Colleges of Technology,
Abu Dhabi, UAE

Romeo-Teodor Cristina
Banat's University of Agricultural Sciences and Veterinary Medicine
"King Michael I of Romania",
Timisoara, Romania

Section 1

Molecular Mechanisms

Chapter 1

Strategic Role Players of Important Antimicrobial-Resistant Pathogens

Shama Mujawar, Bahaa Abdella and Chandrajit Lahiri

Abstract

Over the years, tireless efforts of the concerned scientists have produced various new therapeutics and methods for the treatment of bacterial infections. However, despite the vast regimen of modern antibiotics being corroborated, the diseases caused by the Gram-positive and -negative pathogens has become untreatable, mainly due to the constantly evolving threat of antimicrobial resistance (AMR), thereby leading to huge morbidity and mortality. Moreover, shortage of efficient therapies, lack of successful prevention strategies and availability of only a few effective antibiotics urgently necessitated the development of novel therapeutics and alternative antimicrobial treatments. These developments have been based on the molecular mechanisms of resistance posed by the pathogens during their interactions with the host. Herein, we collate four essential bacterial components like chaperones, efflux pumps, two-component systems and biofilms which can present challenges for the most coveted control of infection. Essentially, we discuss the current knowledge status of these components to provide insight into the complex regulation of virulence and resistance for some medically important multidrug-resistant (MDR) pathogens. This will help the future scientists to clearly focus on some specific proteins to be targeted by against the available class of drugs and/or antibiotics with the broader perspective to develop novel antimicrobial agents.

Keywords: antimicrobial resistance, biofilms, chaperones, efflux pumps, multidrug resistance, two-component systems

1. Introduction

Bacterial infections have been threatening human population since time immemorial. Being one of the leading causes of morbidity and mortality (**Figure 1**), the latest global rise in antibiotic resistance threatens to undo decades of progress in treating such bacterial infectious diseases caused by the pathogens. In fact, multidrug resistance (MDR) conferred by Gram-positive and -negative bacteria is difficult to treat and may even be, untreatable with conventional antibiotics. The case has turned out to be so serious that many of these microorganisms are at least resistant to a single drug regimen while several are moving from developing MDR to extensively and total drug resistance, referred to as XDR and TDR, respectively.

All the aforementioned classes of resistance, namely MDR, XDR and TDR, commonly referred to as antimicrobial resistance (AMR), has been conferred the main cause for the second leading global disease burden of bacterial infection in the twenty-first century, as reported by WHO [1]. Importantly, the development of

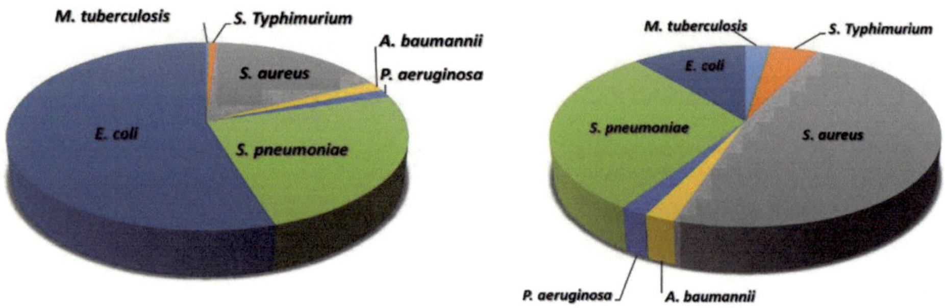

Figure 1.
Rates of infection (left) vs. mortality (right) statistics as per National Center for Health Statistics (CDC), 2017.

antibiotics has directly influenced the initial resistance caused by using newer agents. Moreover, the discovery of new antibiotic classes is reported to be void since 1987, when lipopeptides was the last class introduced (**Figure 2**) [2]. Thus, it has become increasingly difficult to find therapeutic options to treat organisms developing AMR, such as *Acinetobacter baumannii, Proteus mirabilis and Pseudomonas aeruginosa* [2]. Nevertheless, antimicrobials have had a significant positive effect on the administration of irresistible infections and have become a basic component of all perspectives of modern healthcare.

The rise of AMR development has become a serious concern more than what can be even perceived. This is potentiated by different facts ranging from adverse effects of existing antibiotics and consequent re-purposing and/or chemical modification or their withdrawal leading to the sparing usage of new ones due to resistance concerns and ultimately a shortage in the development of new antibiotics [3–5]. Moreover, environments of hospitals and other health care systems as well as social communities and advanced transport systems have enabled the spread of AMR easier and faster [6]. This is evidenced by a recent increase in the carbapenem resistance (e.g. meropenem) due to the presence of carbapenemase, a.k.a. New Delhi

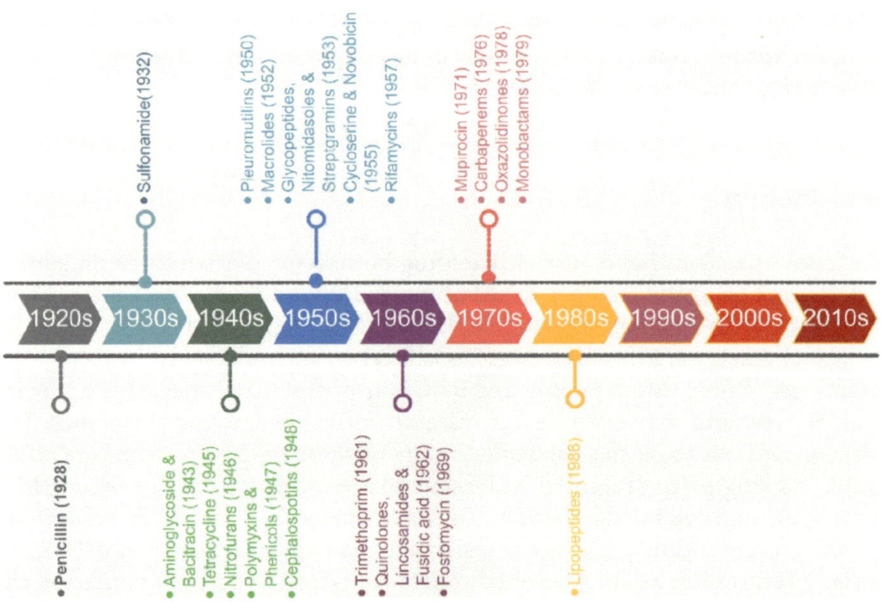

Figure 2.
The timeline of the development of different antibiotic classes.

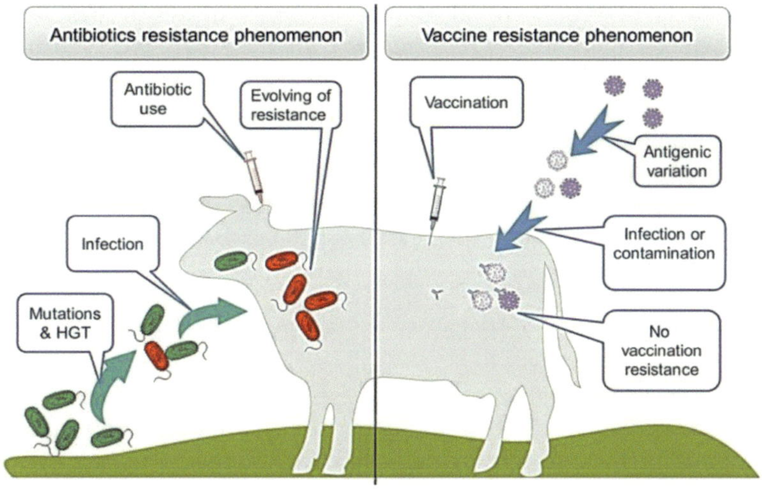

Figure 3.
Antibiotic vs. vaccine-resistant phenomenon. HGT represents horizontal gene transfer, green and red colored cells denote antibiotic-sensitive and -resistant bacteria, respectively.

metallo-β-lactamase-1 (NDM-1) in various Enterobacterales isolates [7]. Initially reported to have found in a patient in Sweden in 2008 who had originated from India, such cases were found later in UK patients having either travel or ancestral history from the Indian subcontinent [6]. A variation of no such travel or hospital contacts, for patients harboring NDM-1, was also reported by 2011 [8], along with drinking water and sewage samples containing a range of NDM-1 harboring bacteria (e.g. *Shigella boydii and Vibrio cholerae*) [9] thereby proving that AMR development varies within organism and with the mechanism of transfer of mobile resistance elements between species (**Figure 3**). Again, some vaccine resistance phenomenon has added on to activities while researchers are aiming to produce advanced vaccines through recombinant DNA technology, keeping in mind the utility of vaccines over antibiotics (**Figure 3**).

2. The causes

AMR is exhibited when a microorganism survives in the presence of an antibiotic concentration that is generally adequate to prevent or stop its growth. Thus, in clinical terms "prone" and "resistant" are generally used to infer the efficacy or failure of medical therapy, respectively [10]. Moreover, the microbes can either be inherently resistant to an antibiotic or develop resistance after their exposure to incorrect and/or insufficient dosage prescription. This is commonly the case for patients routinely communicating with hospital settings thereby having gradually increased resistance to frequently used antibiotics. For these cases of hospital-acquired infections (HAI), certain bacteria develop drug-resistant strains through natural selection mechanism which promotes the persistence of bacterial strains having acquired some mutation [11]. However, the increased profile of these pathogens with AMR varies, even though they arise from similar causes.

AMR resistance may evolve as a mechanistic consequence of gene mutation or direct gene transfer, the latter being also known as horizontal gene transfer (HGT). Of these two, HGT helps to acquire new resistance genes and virulence determinants through a multitude of mechanisms including conjugation, transduction or transformation among related and/or non-related species [10]. This phenomenon is commonly associated with bacterial adaptation to new niches or lifestyles and has an impact on the development of its genomic content. Again, HGT, with the help of mobile genetic elements (MGEs) like transposons, has been reported to have conferred resistance to a broad range of antibiotics, particularly toward new ones. Moreover, transmissible plasmids and phages often bear genes that confer antibiotic resistance to one or more distinct antibiotics and facilitate their transfer across different genera. Such evolution essentially underpins the survival of the developing MDR strains and may be a major reason for the global outbreaks.

3. The effectors

Besides the emerging species of bacteria exhibiting MDR, namely *Salmonella enterica*, *Mycobacterium tuberculosis* and *P. mirabilis*, other common species of MDR bacteria responsible for two-thirds of all HAIs are defined by the acronym ESKAPE to denote the six pathogens namely, *Enterococcus faecium, S. aureus, Klebsiella pneumoniae, A. baumannii, P. aeruginosa* and *Enterobacter* spp. [12]. These are easily distinguishable from other pathogens due to their enhanced resistance to frequently used antibiotics such as penicillin, vancomycin, carbapenems and more. One of the common resistance mechanisms involves enzyme production that alters the antibiotic target sites and results in no binding activity with efflux pumps [13]. Efflux pumps are the characteristics of the Gram-negative bacterial membrane that enables them to constantly pump out foreign materials, including antibiotics, such that the intracellular milieu does not have sufficiently elevated drug concentration to make the effect [13]. Moreover, biofilms are a combination of different microbial and polymer groups that protect the bacteria from antibiotic therapy by acting as a biological barrier [13].

3.1 *Salmonella enterica*

Human infections due to *S. enterica*, a bacterial pathogen, constitute significant food borne disease burdens of blood stream associated with a high mortality ratio throughout the world [14, 15]. *S. enterica* are the Gram-negative facultative anaerobe that belongs to the family Enterobacteriaceae. From over 2,500 strain types, the strain *S. enterica serovar* Typhi causes the typhoid fever [16]. Infections with *Salmonella* in humans typically range from non-typhoidal salmonella (NTS) to typhoidal fever, which can be life-threatening. Additionally, the resistant serovars causing enteric fever, namely, Typhi, Paratyphi A, B, or C are broadly referred to as typhoidal *Salmonella* serovars [14]. However, these are highly adapted to the human host that is used as their exclusive reservoir [17].

The initial AMR acquired by *Salmonella* was to the first-line drugs such as ampicillin, chloramphenicol and sulfamethoxazole. The AMR mechanisms in *S.* Typhi include drug inactivation, target site modification and active efflux, which might be chromosomal or plasmid-mediated [18]. In fact, the resistance of *Salmonella* and pathogenic *E. coli* along with other Gram-negative bacteria, against antibiotic and non-antibiotic compounds, is related to efficient efflux pumps, which reduces the intracellular concentration of such compounds [19, 20]. The occurrence of plasmid-mediated antibiotic resistance to fluoroquinolones has recently been recorded and referred to a single point mutation in the topoisomerase gene gyrA, encoding DNA gyrase. Moreover, pathogenic *Salmonella* uses the two-component systems (TCS) namely, PhoPQ, PmrAB and Rcs regulatory system for lipopolysaccharide (LPS) modification and increases the resistance toward host human AMPs [18], which could help it to survive *in vivo* and develop the disease [21]. The lack of such systems in the eukaryotic host made them eligible to be targeted by anti-virulence compounds. This strategy was rendered successful during selective active site inhibition of PhoQ autokinase activity by Quinazoline [22]. Furthermore, for cells lacking an RNA chaperone, known as bacterial cold shock proteins (CSPs), high levels of porin genes *viz. omp*D, *omp*F, and *omp*C resulted in increased cell membrane permeability in response to bile salt stress [23]. This finding highlights on the importance of the chaperone protein in the maintenance of the membrane integrity and selective permeability.

3.2 *Proteus mirabilis*

P. mirabilis, the Gram-negative uropathogenic bacteria and a member of the Enterobacteriaceae family, is developing MDR to antibiotics and biofilm formation. This may trigger significant complications in patients with long-term catheters or urinary tract infections (UTI) [24, 25]. *Salmonella* genomic island 1, an integrative mobilizable component of multidrug-resistant *S.* Typhimurium, was recently identified in a remarkably high proportion of *P. mirabilis* clinical isolates from France, indicating its involvement in the spread of this MDR element [26]. *P. mirabilis* is susceptible to aminoglycosides, fluoroquinolones, sulfamethoxazole and β-lactams, but resistant to tetracycline and nitrofurantoin [27]. This enhanced resistance to antimicrobial agents has resulted not only in modifications to antimicrobial therapies, but also in poor diagnosis and increased mortality rate of nosocomial infections [28]. Astonishingly, in 2016, a new isolate of *P. mirabilis*, from diabetic ulcer patient, have shown a remarkable resistance to silver nanoparticles, spreading the alarm of resistance to even include metallic nanoparticles like silver [29]. Moreover, the efflux pumps (EPs) also play important role in *P. mirabilis* drug resistance, as exemplified of the increased cell permeability of the EP inhibitor Phenylalanine-Arginine Beta-Naphthylamide (PAβN), thereby making it more susceptible to acetylsalicylic acid [30].

3.3 *Acinetobacter baumannii*

A. baumannii are the most successful Gram-negative opportunistic nosocomial pathogens responsible for hospital-acquired infections (HAI) in intensive care units. The WHO has stated *A. baumannii* to be one of the most serious ESKAPE organisms that have effectively escaped the effects of antibiotics [31]. Several resistance mechanisms are known, including target modifications, multidrug efflux pumps, enzymatic degradation or modification of drugs and permeability defects besides some other uncategorized ones [31]. *A. baumannii* strain isolates are resistant to cephalosporins and penicillins, including inhibitory combinations of aminoglycosides and fluoroquinolones [32]. Moreover, some *A. baumannii* strains can acquire families of EP from other species and new β-lactamases to improve the resistance of β-lactam antibiotics [11]. Furthermore, *A. baumannii* clinical isolates are reported to be resistant to colistin developed due to a modification of the lipid A component of the lipopolysaccharide outer membrane. The modification is mediated by the TCS PmrAB and a mutation of the LpxA/C/D gene [33]. Such resistance mechanism through LPS modification brings out the importance of the TCS in *A. baumannii*.

3.4 *Staphylococcus aureus*

S. aureus is a major Gram-positive pathogen, both within the hospital settings and environmental communities and reported to be prone to nearly any antibiotic ever produced [34]. Such multiple antibiotics resistance has developed by acquiring MGE through HGT. This results in mutations that alter drug binding sites on molecular targets leading to an increase in the expression of endogenous efflux pumps. These resistant strains fight antibiotics by deactivating β-lactam binding proteins [11]. Due to its increasing antibiotic resistance to penicillin and methicillin, the bacteria remain a growing pandemic through mechanisms including HGT and antibiotic alterations [35]. Moreover, *S. aureus* is not far from the Gram-negative bacteria which are resistant to antibiotics mediated by TCS. Thus, the TCS $VanR_AS_A$ regulates the necessary mechanism of resistance in vancomycin resistance *S. aureus* (VRSA) [36]. Again, the EPs from *S. aureus* have been categorized recently in six different diverse groups. They were found to be either chromosomal or extrachromosomal except *qac*A/B and *smr* which were found only on the studied plasmid samples [37].

3.5 *Pseudomonas aeruginosa*

P. aeruginosa, the Gram-negative nosocomial pathogen, is considered as an epitome of AMR due to its major involvement in causing chronic and nosocomial diseases. This high rate of resistance is directly related to their various inherent resistance mechanisms expressed, including the down-regulation of porin manufacturing system (carbapenems and cefepime), overexpression of efflux pumps (carbapenems) or production of other beta-lactamases besides the high production of AmpC beta-lactamase. The most frequently administered antipseudomonal antibiotics are aminoglycosides, fluoroquinolones and β-lactams that are susceptible to the known resistance mechanisms in *P. aeruginosa*. Its mutants, with upregulated EPs, have been reported that makes it difficult to find an effective antibiotic [38]. Moreover, only inhibition of the EP, in the recent clinical MDR isolates, has almost no effect in increasing susceptibility toward the tested antibiotics [39]. However, inhibition of histidine kinases (HKs), a part of TCS, using benzothiazole-based HK inhibitors, resulted in a reduced production of molecules which are linked to quorum-sensing and redox-balance. It also showed reduced motility and attachment ability, rendering it to be less virulence [40].

3.6 *Mycobacterium tuberculosis*

Tuberculosis (TB) poses serious global health crisis as an important chronic infectious disease caused by strains of *M. tuberculosis* (MTB). It is an extremely dangerous human pathogen that infects one-third of the world's population and causes almost two million fatalities each year [41]. Besides that, the total number of cases have been still increasing, due to strains of MTB being resistant to first-line drug therapy [41]. This involves resistance to the two most powerful anti-TB drugs, rifampicin and isoniazid, thereby evoking the title of multidrug resistance TB (MDR-TB). The existence of even more resistant MTB strains has been described as extreme drug-resistant (XDR)-TB, which shows resistance against the injectable second line drugs such as kanamycin, amikacin or capreomycin [42]. A more alarming situation has arisen with the depiction of MTB strains showing resistance to all antibiotics available for testing, with the species being termed as total drug resistant (TDR)-TB [43]. Therefore, the early onset of detection and prevention of MDR-TB, will enable the therapeutic treatment to reduce the spread of infection. Thus, a better understanding of the mechanisms of action of anti-TB drugs will facilitate the development of new drug targets aimed at improving outcomes from diseased patients [44]. In fact, it has been found that MprA, part of TCS MprAB.

along with other TCS, namely, TrcRS, control the expression of antibiotic resistant related β-propeller gene Rv1057 [45, 46]. The roles of chaperone(s) in the resistance of *M. tuberculosis* are yet to be declared.

3.7 *Klebsiella pneumoniae*

K. pneumoniae is a Gram-negative hospital-acquired pathogen causing nosocomial pneumonia and urinary tract infections. The increased incidence of carbapenemase-producing and thus, carbapenem resistant *K. pneumoniae* (CRKP), has posed a major threat to global human health. Diseases caused by CRKP were treated successfully in combination therapies of antimicrobial agents [47]. It has been reported that tigecycline and the polymyxins (polymyxin B or colistin) showed variable susceptibilities to treat infections caused by CRKP [4]. This has led to the emergence of CRKP, against which there are very few antibiotics in development that can treat the infections [11]. Incidentally, the minimum inhibitory concentration (MIC) of eravacycline has been increased as a consequence of increased expression of two EP complexes OqxAB and MacAB in *K. pneumoniae* [48], which suggests their contribution to resistance against this antibiotic.

3.8 *Enterococcus faecium*

Generally associated with HAI in immunocompromised patients, *E. faecium* is a Gram-positive bacterium, often showing resistance to β-lactam antibiotics, including penicillin and other antibiotics of last resort. Reportedly, there has also been an increase in vancomycin-resistant enterococci (VRE) strains, exhibiting resistance to vancomycin-A [11]. These VRE strains show an ability to produce and share their resistance through HGT, as well as code for virulence factors that regulate phenotypes. These virulence phenotypes differ from the wild types in producing thicker biofilms for development in a variety of environments, including medical devices such as urinary catheters and cardiovascular prosthetic valves [14]. The thicker biofilms function as a "mechanical and biochemical shield" that protects the bacteria from antibiotics and is the most efficient protective system against bacterial treatment [13]. In fact, the intensive use of antibiotics in animal rearing resulted in the development of resistance in *E. faecium* [49]. Moreover, recently, a study identified few new antibiotics resistance genes related to EP, namely, *optrA* and *poxtA* besides the new gene *cfr*-like variant in *E. faecium* [50]. Earlier, the expression of the EP proteins, EfrAB, has been shown to be increased upon halving the MIC of gentamicin and got lowered upon the addition of 3 mM EDTA [51]. Furthermore, TCS like ChtSR has been found to be responsible for chlorhexidine tolerance in MDR *E. faecium*, upon testing by targeted deletion mutation of *cht*R and *cht*S genes [52]. Again, the TCS CroRS was reported to be crucial in resistance to cell wall antibiotics in *E. faecium* [53].

3.9 *Enterobacter*

Enterobacter are Gram-negative bacterial species which trigger UTI and blood diseases. They show resistance against different drug therapies, thus, requiring the development of new and efficient antibiotic treatments [54]. In fact, colistin and tigecycline are, only two of the antibiotics, presently being used as medication while no other feasible antibiotics are apparently being developed. Other most commonly reported antimicrobials in *Enterobacter* infections are aminoglycosides, cephalosporins, carbapenems and fluoroquinolones. Moreover, in some species, a 5- to 300-fold rise in the MIC was reported when subjected to several gradually increasing benzalkonium chloride concentrations [55]. In fact, the EP protein, SugE, in *Enterobacter cloacae,* which is a member of small multidrug resistance (SMR) protein family, has been found to be responsible for resistance against toxic compounds such as cetylpyridinium chloride and benzalkonium chloride [56]. Another EP protein of the resistance nodulation cell division (RND) category, AcrB, has also been found to be very essential in the pathogenicity and antibiotics resistance of *E. cloacae* [57, 58].

4. The role players

Several health interventions have been proposed as alternatives to current antibiotic therapy and prevent the resistant mechanisms of which, the development of new drug classes, use of vaccines or other therapeutic strategies are noteworthy (**Figure 4**) [59]. In fact, using computational approaches, certain proteins and/or phenotypes, having plausible involvement in antibiotic resistance, are proposed [60–64] as discussed below.

4.1 Chaperones

Bacterial chaperones like DnaK, belonging to the heat shock proteins (Hsp)70 family, are produced by cells in response to exposure to stressful conditions [65]. The interaction between the two domains of such Hsp70, namely ATPase and the substrate-binding domain, triggers the chaperone-based activity of DnaK which are also enhanced by the co-chaperone such as DnaJ (Hsp40 family) and chaperone GroE (Hsp60 family) [66]. DnaK acts on unfolded and partially folded protein chains by binding and controlling their configuration [67].

Besides stress response, DnaK plays a significant housekeeping role in maintaining normal bacterial cellular growth and homeostasis [68]. Thus, any

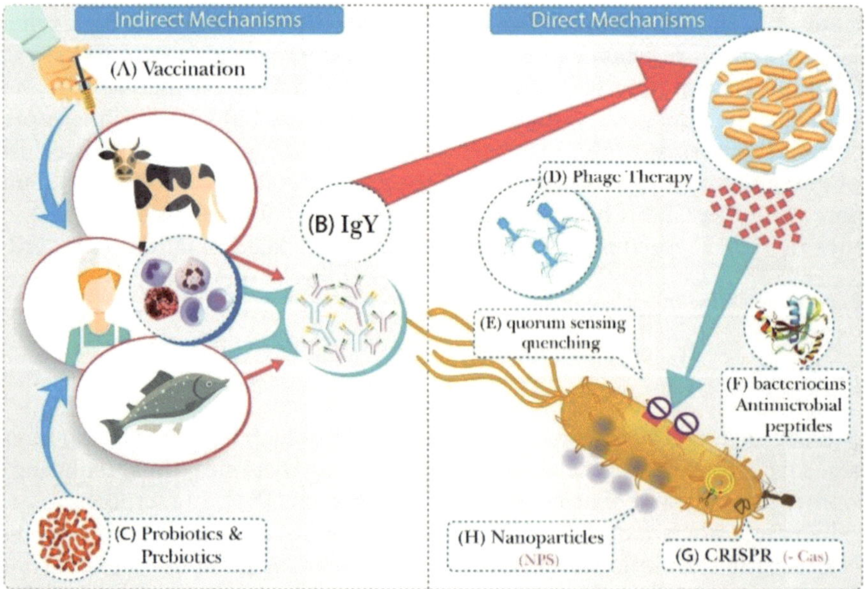

Figure 4.
Alternative strategies to combat antimicrobial resistance and their mechanisms of action.

alterations in the *dnaK* gene reduce the growth of bacteria within the host [69]. In fact, during infection, bacteria activate their heat shock genes like DnaK to protect their cellular machinery from the consequently activated host immune system for defense mechanisms and thereby strengthen their virulence strategy [69]. This phenomenon, thus, provides an insight into structural mechanism of DnaK, leading to misfolding and its role in controlling protein activity contributing to the pathogenicity of multidrug-resistant bacteria, such as the opportunistic human pathogen *A. baumannii* [70]. In fact, DnaK mutants showed decreased viability and improved susceptibility under strained circumstances during systemic infection as reported for *dnaK* mutants of *S. aureus* with increased sensitivity to oxacillin and methicillin [71] and *dnaK/dnaJ* mutants of *E. coli* having increased sensitivity to fluoroquinolones [72].

4.2 Efflux pumps

Antibiotic resistance can be triggered, in MDR bacteria, by four discrete mechanisms viz. target modification, reduced permeability and improved efflux, drug inactivation and drug extrusion by the multidrug efflux pumps (EP) [73]. Due to their poly-substrate specificity, besides having the potential to expel a broad variety of antibiotics, these EP also manage the development of other resistance mechanisms by decreasing intracellular antibiotics concentration and stimulating mutation accumulation [73]. Consequently, over-expression of multidrug EP is involved with clinically related antibiotic resistance. Thus, there has been increasing evidence of EP having biochemical functions in bacteria along with their appearance under strict regulations in response to some physiological and environmental signals [73]. Hence, a systematic knowledge of EP is important for the development of EP inhibitors as promising AMR intervention strategies.

EP are present in almost all bacterial species involved in AMR. They can be located on plasmids or chromosomes that encode this class of proteins. The five families of bacterial EP, found to be involved in MDR, are the major facilitator superfamily (MFS), the multidrug and toxic compound extrusion (MATE) family, the ATP-binding cassette (ABC) superfamily, the small multidrug resistance (SMR) family, and the resistance-nodulation-division (RND) family, based on their composition, energy sources and substrates used [73]. Importantly, only RND superfamily is found in Gram-negative bacteria due to its structure containing tripartite complex and the efflux systems of the other four families are widely distributed in both Gram-positive and -negative bacteria. These EP can be either single or multiple-component transporters depending on their specific classes. They comprise both an inner and an outer membrane transporter, like the RND type. It has been found that RND family pumps are frequently associated with therapeutically important bacterial resistance such as AcrB in *S.* Typhimurium and *E. coli* and MexB in *P. aeruginosa* owing to their tripartite complex, enabling various drugs to be immediately extruded from cytoplasm to outside the bacterial cells [74].

In fact, antibiotics such as fluoroquinolone, tetracycline, rifampin, novobiocin, chloramphenicol and B-lactams were used to analyze the substrate profile of housekeeping efflux system AcrAB-TolC in *E. coli* [74]. Similarly, the *S.* Typhimurium AcrAB-TolC efflux system was also capable of expelling various antibacterial agents such as tetracycline, quinolones and chloramphenicol [75, 76]. The two RND efflux pumps, MexAB-OprM and MexXY-OprM, homolog to AcrAB-TolC system in *E. coli*, are also expressed in *P. aeruginosa*. Thus, these systems can actively export chloramphenicol, tetracycline and fluoroquinolones. In addition to these substrates, MexAB-OprM export B-lactams and novobiocin whereas MexXY system exports aminoglycosides (**Table 1**) [77].

Efflux pump	Pump type	Regulator	Regulator family	Inducible signal
Acinetobacter baumannii				
AdeABC	RND	AdeRS	TCS	~
Pseudomonas aeruginosa				
MexXY	RND	MexZ	TetR	Tetracycline, erythromycin, gentamicin
MexAB	RND	MexR	MarR	Superoxide stress
		NalD	TetR	~
MexCD	RND	NfxB	LacI/GalR	Biocide chlorhexidine
MexEF	RND	MetT	LysR	Chloramphenicol, GSNO
Salmonella Typhimurium				
AcrAB	RND	MarA	AraC	~
		RamA	AraC	Indole, bile salts
		RamR	TetR	~
		SoxS	AraC	~
		AcrR	TetR	~
AcrD	RND	BaeSR	TCS	Indole, zinc, copper
		CpxAR	TCS	Indole, zinc, copper
AcrEF	RND	AcrS	TetR	~
MdtABC	RND	BaeSR	TCS	Indole, zinc, copper
		CpxAR	TCS	Indole, zinc, copper
MacAB	ABC	PhoQP	TCS	Magnesium
MdsABC	RND	GolS	MerR	Gold
Staphylococcus aureus				
MepA	MFS	QacR	TetR	Rhodamine 6G, TPP
QacA	MATE	MepR		Chlorhexidine, cetrimide, dequalinium

Adapted from [73].
~ means unknown.

Table 1.
Selected multidrug resistance efflux pump regulators.

4.3 Two-component systems

Two component systems (TCS) are commonly found in bacteria, allowing them to respond to various fluctuations in the environment. Canonically, TCS are composed of a response regulator (RR) and a histidine kinase (HK) [78]. The membrane associated HKs can detect and transform various environmental sensations by autophosphorylation. The HKs can then transphosphorylate their cognate partners, the RRs, which then influence the expression of downstream genes to affect the concerned phenotype [78].

A thorough investigation of the correlation between efflux pumps and TCSs in *E. coli*, revealed the involvement of several RRs in drug resistance [79]. Among them, *mdtABC* and *acrD* expression was triggered by the BaeSR and CpxAR TCS in response to indole [80, 81] and envelope stress [79], respectively, while no signals were detected when the EvgSA TCS triggered the activation of EmrKY and MdtEF [82, 83]. Moreover, the expression of the MdtEF efflux pump was

Bacteria	Inhibitors	TCS	Mechanisms	Reference
Salmonella enterica Typhi and/or Typhimurium	XR770	BaeSR OmpR/ EnvZ	Inhibition of key interacting residues of DHp domain of HK	[89]
	NSC9608 (8 compounds, NCI library)	PhoP-Q	Inhibition of formation of the PhoP-DNA complex	[90]
Pseudomonas aeruginosa	Thiazole derivatives	Algr1-2	Inhibition of phosphorylation/ dephosphorylation of Algr2 Inhibition of DNA-binding activity of Algr1	[90]
Enterococcus faecium	Thiazole derivatives	VanR-S	Inhibition of autophosphorylation	[90]
Staphylococcus aureus	Walkmycin B and Waldiomycin	WalK-R	Binds to the HK cytoplasmic domain for the inhibition of autophosphorylation	[90]
	Salicylanilide	KinA/ Spo0F	Affects membrane fluidity, disturbing signal transduction	[90]
Methicillin-resistant Staphylococcus aureus	Bis-phenol	VanR-S	—	[90]

Table 2.
Representative TCS targets with their known inhibitors.

triggered by ArcAB-TCS system in the M9 glucose medium [84, 85]. Similar to *E. coli*, MdtABC and AcrD are also stimulated by the *Salmonella* BaeSR TCS in response to metal ions [86].

Again, PhoPQ TCS, the core virulence regulator in *Salmonella*, controls the activation of the RND type MacAB pump [87, 88]. TCS was also revealed to be involved in regulating efflux pumps in other species. In *A. baumannii*, the expression of RND type efflux pump AdeABC has been reported to be regulated by AdeRS-TCS (**Table 2**). The AdeRS-TCS regulatory system is encoded by *adeRS* genes, being positioned in the upstream region of *adeABC* genes [91]. Inactivation of AdeR or AdeS resulted in *A. baumannii* being susceptible to aminoglycosides which are the substrates of this pump, indicating the vital role of AdeRS in *adeABC* activity. The nature of these inducing signals and the AdeRS activation mechanism, however, remain unclear [92].

TCS can play an important role in drug discovery. There are several ways to target TCS proteins. Of these, the structure-based virtual screening (SBVS) analysis is carried out using compound databases containing a broad range of prospective inhibitors, including structures known to be antibacterial [93].

4.4 Biofilms

Biofilms, in both single and multi-species groups, communicate and cooperate to perform complex processes with each other and their environment [94]. With the scientists aiming to understand the intercellular interactions that encourage the development of biofilms, they are presently a serious health issue, playing a major role in abiotic device-related diseases such as catheters, prosthetic valves and contact lenses [95].

Biofilm formation can be explored in different stages comprising (a) distinctive adhesion of the planktonic bacteria (PB) to a solid surface [96], (b) micro-colonies

(MC) formation surrounded by protective secreted molecules known as the matrix of extra polymeric substances (EPS) having up to 97% water as the main component [97] and (c) dispersal including shedding of PB or MC from the mature biofilm [97]. The last phase can encourage further biofilm colonization of the host which can eventually benefit the bacteria with a limited supply of nutrients and waste accumulation [97]. Importantly, the transition from planktonic growth to surface life is triggered by several environmental signals known as various stresses for the bacteria based on their ecological niche [98]. These include UV radiation, pH changes, oxygen tension, osmolarity, iron availability, temperature, nutrient supply and desiccation [98], which may disrupt their fundamental functions such as growth and survival capability. The environmental indications, however, vary significantly between organisms. Thus, *P. aeruginosa* will form biofilms under most circumstances [99, 100] while *E. coli* O157:H7 produce a biofilm under low-nutrient conditions only [101].

Recent advances in biofilm research have proven its connection to various pathways and proteins [61]. For instance, defects in MDR EP activity reduced the biofilm formation and thus, EP inhibitors have been employed as a promising biofilm inhibition approach for strains of *E. coli* and *Klebsiella* [102], *Salmonella* [103], *P. aeruginosa* and *S. aureus* [104]. However, certain other reports show that despite the elimination of planktonic cells through pharmacological intervention, the sessile forms are resistant and continue to proliferate within the biofilm [105]. This is more of prominence on abiotic surfaces [95], such as catheters [106], contact lenses [107] and prosthetic cardiac valves [108]. Thus, alginate mucoids, with EPS overexpression, from *P. aeruginosa* species isolated from cystic fibrosis patients, were found to improve AMR by promoting the biofilm formation [109].

5. Conclusion

The constant increase in AMR is a significant public health concern that needs to be addressed now. This review starts with an introduction to AMR followed by the threats from the clinically important MDR pathogens and their rise. With the existing management strategies for MDR by the scientists still ongoing, we have taken up this study to propose an integrated approach to deal with MDR threats. Thus, the review ends with new connections of important bacterial components with MDR.

Acknowledgements

The authors acknowledge the support of Sunway University, Selangor, Malaysia for providing the computational facilities and wishes to thank Hend Salah for the valuable contribution in developing the artwork for the concept provided.

Conflict of interest

The authors declare no conflict of interest.

Author details

Shama Mujawar[1]†, Bahaa Abdella[1,2]† and Chandrajit Lahiri[1]*

1 Department of Biological Sciences, Sunway University, Petaling Jaya, Selangor, Malaysia

2 Faculty of Aquatic and Fisheries Sciences, Kafrelsheikh University, Egypt

*Address all correspondence to: chandrajitl@sunway.edu.my

† These authors have contributed equally to the work.

References

[1] Conly J, Johnston B. Where are all the new antibiotics? The new antibiotic paradox. Canadian Journal of Infectious Diseases and Medical Microbiology. 2005;**16**:159-160. DOI: 10.1155/2005/892058

[2] Livermore DM. Has the era of untreatable infections arrived? Journal of Antimicrobial Chemotherapy. 2009; **64**:i29-i36. DOI: 10.1093/jac/dkp255

[3] Frieden T. Antibiotic Resistance Threats in the United States. USA: Centers for Disease Control and Prevention; 2013. Available from: https://www.cdc.gov/drugresistance/pdf/ar-threats-2013-508.pdf

[4] Gilbert DN, Guidos RJ, Boucher HW, Talbot GH, Spellberg B, Edwards JE, et al. The 10 × '20 initiative: Pursuing a global commitment to develop 10 new antibacterial drugs by 2020. Clinical Infectious Diseases. 2010;**50**:1081-1083. DOI: 10.1086/652237

[5] Sabtu N, Enoch DA, Brown NM. Antibiotic resistance: What, why, where, when and how? British Medical Bulletin. 2015:**111**:105-113. DOI: 10.1093/bmb/ldv041

[6] Rooney PJ, O'Leary MC, Loughrey AC, McCalmont M, Smyth B, Donaghy P, et al. Nursing homes as a reservoir of extended-spectrum-lactamase (ESBL)-producing ciprofloxacin-resistant *Escherichia coli*. The Journal of Antimicrobial Chemotherapy. 2009;**64**:635-641. DOI: 10.1093/jac/dkp220

[7] Adeolu M, Alnajar S, Naushad S, Gupta RS. Genome-based phylogeny and taxonomy of the 'Enterobacteriales': Proposal for Enterobacterales ord. nov. divided into the families Enterobacteriaceae, Erwiniaceae fam. nov., Pectobacteriaceae fam. nov., Yersiniaceae fam. nov., Hafniaceae fam. nov., Morganellaceae fam. nov., and Budviciaceae fam. nov. International Journal of Systematic and Evolutionary Microbiology. 2016;**66**:5575-5599. DOI: 10.1099/ijsem.0.001485

[8] Kumarasamy KK, Toleman MA, Walsh TR, Bagaria J, Butt F, Balakrishnan R, et al. Emergence of a new antibiotic resistance mechanism in India, Pakistan, and the UK: A molecular, biological, and epidemiological study. The Lancet Infectious Diseases. 2010;**10**:597-602. DOI: 10.1016/S1473-3099(10)70143-2

[9] Woodford N, Johnson AP. Global spread of antibiotic resistance: The example of New Delhi metallo-β-lactamase (NDM)-mediated carbapenem resistance. Journal of Medical Microbiology. 2013;**62**:499-513. DOI: 10.1099/jmm.0.052555-0

[10] Livermore D. Can better prescribing turn the tide of resistance? Nature Reviews. Microbiology. 2004;**2**:73-78. DOI: 10.1038/nrmicro798

[11] Pendleton JN, Gorman SP, Gilmore BF. Clinical relevance of the ESKAPE pathogens. Expert Review of Anti-Infective Therapy. 2013;**11**(3):297-308. DOI: 10.1586/eri.13.12

[12] Mulani MS, Kamble EE, Kumkar SN, Tawre MS, Pardesi KR. Emerging strategies to combat ESKAPE pathogens in the era of antimicrobial resistance: A review. Frontiers in Microbiology. 2019;**10**:539. DOI: 10.3389/fmicb.2019.00539

[13] Santajit S, Indrawattana N. Mechanisms of antimicrobial resistance in ESKAPE pathogens. BioMed Research International. 2016;**2016**:2475067. DOI: 10.1155/2016/2475067

[14] Gal-Mor O, Boyle EC, Grassl GA. Same species, different diseases: How

and why typhoidal and non-typhoidal *Salmonella enterica* serovars differ. Frontiers in Microbiology. 2014;**5**:391. DOI: 10.3389/fmicb.2014.00391

[15] Reddy EA, Shaw AV, Crump JA. Community-acquired bloodstream infections in Africa: A systematic review and meta-analysis. The Lancet Infectious Diseases. 2010;**10**(6):417-432. DOI: 10.1016/S1473-3099(10)70072-4

[16] Johnson R, Ravenhall M, Pickard D, Dougan G, Byrne A, Frankel G. Comparison of *Salmonella enterica* serovars Typhi and Typhimurium reveals typhoidal serovar-specific responses to bile. Infection and Immunity. 2018;**86**(3): e00490-17. DOI: 10.1128/IAI.00490-17

[17] Deen J, von Seidlein L, Andersen F, Elle N, White NJ, Lubell Y. Community-acquired bacterial bloodstream infections in developing countries in south and Southeast Asia: A systematic review. The Lancet Infectious Diseases. 2012;**12**:480-487. DOI: 10.1016/S1473-3099(12)70028-2

[18] Feasey NA, Dougan G, Kingsley RA, Heyderman RS, Gordon MA. Invasive non-typhoidal salmonella disease: An emerging and neglected tropical disease in Africa. The Lancet. 2012;**379**: 2489-2499. DOI: 10.1016/S0140-6736(11)61752-2

[19] Laudy AE. Non-antibiotics, efflux pumps and drug resistance of Gram-negative rods. Polish Journal of Microbiology. 2018;**67**:129-135. DOI: 10.21307/pjm-2018-017

[20] Weston N, Sharma P, Ricci V, Piddock LJV. Regulation of the AcrAB-TolC efflux pump in Enterobacteriaceae. Research in Microbiology. 2018;**169**:425-431. DOI: 10.1016/j.resmic.2017.10.005

[21] Andersson DI, Hughes D, Kubicek-Sutherland JZ. Mechanisms and consequences of bacterial resistance to antimicrobial peptides. Drug Resistance Updates. 2016;**26**:43-57. DOI: 10.1016/j.drup.2016.04.002

[22] Carabajal MA, Asquith CRM, Laitinen T, Tizzard GJ, Yim L, Rial A, et al. Quinazoline-based antivirulence compounds selectively target *Salmonella* PhoP/PhoQ signal transduction system. Antimicrobial Agents and Chemotherapy. 2019;**64**:e01744-19. DOI: 10.1128/AAC.01744-19

[23] Ray S, Da Costa R, Das M, Nandi D. Interplay of cold shock protein E with an uncharacterized protein, YciF, lowers porin expression and enhances bile resistance in *Salmonella* Typhimurium. The Journal of Biological Chemistry. 2019;**294**:9084-9099. DOI: 10.1074/jbc.RA119.008209

[24] Hall RM, Collis CM. Antibiotic resistance in Gram-negative bacteria: The role of gene cassettes and integrons. Drug Resistance Updates. 1998;**1**(2): 109-119. DOI: 10.1016/S1368-7646(98)80026-5

[25] Endimiani A, Luzzaro F, Brigante G, Perilli M, Lombardi G, Amicosante G, et al. *Proteus mirabilis* bloodstream infections: Risk factors and treatment outcome related to the expression of extended-spectrum β-lactamases. Antimicrobial Agents and Chemotherapy. 2005;**49**(7):2598-2605. DOI: 10.1128/AAC.49.7.2598-2605.2005

[26] Doublet B, Poirel L, Praud K, Nordmann P, Cloeckaert A. European clinical isolate of *Proteus mirabilis* harbouring the Salmonella genomic island 1 variant SGI1-O. Journal of Antimicrobial Chemotherapy. 2010;**65**(10):2260-2262. DOI: 10.1093/jac/dkq283

[27] Thornsberry C, Yee YC. Comparative activity of eight antimicrobial agents against clinical bacterial isolates from the United States, measured by two methods. The

American Journal of Medicine. 1996;**100** (6A):26S-38S. DOI: 10.1016/s0002-9343 (96)00105-2

[28] Giamarellos-Bourboulis EJ, Papadimitriou E, Galanakis N, Antonopoulou A, Tsaganos T, Kanellakopoulou K, et al. Multidrug resistance to antimicrobials as a predominant factor influencing patient survival. International Journal of Antimicrobial Agents. 2006;**27**(6): 476-481. DOI: 10.1016/j.ijantimicag. 2005.12.013

[29] Saeb ATM, Al-Rubeaan KA, Abouelhoda M, Selvaraju M, Tayeb HT. Genome sequencing and analysis of the first spontaneous nanosilver resistant bacterium *Proteus mirabilis* strain SCDR1. Antimicrobial Resistance and Infection Control. 2017;**6**:119. DOI: 10.1186/s13756-017-0277-x

[30] Laudy AE, Mrowka A, Krajewska J, Tyski S. The influence of efflux pump inhibitors on the activity of non-antibiotic NSAIDS against Gram-negative rods. PLoS One. 2016;**11**: e0147131. DOI: 10.1371/journal. pone.0147131

[31] Lee CR, Lee JH, Park M, Park KS, Bae IK, Kim YB, et al. Biology of *Acinetobacter baumannii*: Pathogenesis, antibiotic resistance mechanisms, and prospective treatment options. Frontiers in Cellular and Infection Microbiology. 2017;7:55. DOI: 10.3389/ fcimb.2017.00055

[32] Greene C, Vadlamudi G, Newton D, Foxman B, Xi C. The influence of biofilm formation and multidrug resistance on environmental survival of clinical and environmental isolates of *Acinetobacter baumannii*. American Journal of Infection Control. 2016;**44**(5):e65-e71. DOI: 10.1016/j.ajic.2015.12.012

[33] Zhang W, Aurosree B, Gopalakrishnan B, Balada-Llasat J-M, Pancholi V, Pancholi P. The role of LpxA/C/D and pmrA/B g ene systems in colistin-resistant clinical strains of *Acinetobacter baumannii*. Frontiers in Laboratory Medicine. 2017;**1**:86-91. DOI: 10.1016/j.flm.2017.07.001

[34] Chatterjee I, Becker P, Grundmeier M, Bischoff M, Somerville GA, Peters G, et al. *Staphylococcus aureus* ClpC is required for stress resistance, aconitase activity, growth recovery, and death. Journal of Bacteriology. 2005;**187**:4488-4496. DOI: 10.1128/JB.187.13.4488-4496.2005

[35] McDonald M, Dougall A, Holt D, Huygens F, Oppedisano F, Giffard PM, et al. Use of a single-nucleotide polymorphism genotyping system to demonstrate the unique epidemiology of methicillin-resistant *Staphylococcus aureus* in remote aboriginal communities. Journal of Clinical Microbiology. 2006;**44**:3720-3727. DOI: 10.1128/JCM.00836-06

[36] McGuinness WA, Malachowa N, DeLeo FR. Vancomycin resistance in *Staphylococcus aureus*. The Yale Journal of Biology and Medicine. 2017;**90**: 269-281. Available from: http://www. ncbi.nlm.nih.gov/pubmed/28656013

[37] Hassanzadeh S, Ganjloo S, Pourmand MR, Mashhadi R, Ghazvini K. Epidemiology of efflux pumps genes mediating resistance among *Staphylococcus aureus*: A systematic review. Microbial Pathogenesis. 2020;**139**:103850. DOI: 10.1016/j.micpath.2019.103850

[38] Levy SB, Bonnie M. Antibacterial resistance worldwide: Causes, challenges and responses. Nature Medicine. 2004;**10**(12 Suppl):S122-S129. DOI: 10.1038/nm1145

[39] Cunrath O, Meinel DM, Maturana P, Fanous J, Buyck JM, Auguste PS, et al. Quantitative contribution of efflux to multi-drug resistance of clinical *Escherichia coli* and

Pseudomonas aeruginosa strains. eBioMedicine. 2019;**41**:479-487. DOI: 10.1016/j.ebiom.2019.02.061

[40] Goswami M, Espinasse A, Carlson EE. Disarming the virulence arsenal of *Pseudomonas aeruginosa* by blocking two-component system signaling. Chemical Science. 2018;**9**: 7332-7337. DOI: 10.1039/c8sc02496k

[41] Palomino JC, Martin A. Drug resistance mechanisms in *Mycobacterium tuberculosis*. Antibiotics. 2014;**3**(3):317-340. DOI: 10.3390/antibiotics3030317

[42] Wright A, Bai G, Barrera L, Boulahbal F, Gilpin C, Drobniewski F, et al. Emergence of *Mycobacterium tuberculosis* with extensive resistance to second-line drugs. Morbidity and Mortality Weekly Report. Annals of Pharmacotherapy. 2006;**40**:1007-1008. DOI: 10.1345/aph.1N108. Available from: https://www.cdc.gov/mmwr/preview/mmwrhtml/mm5511a2.htm

[43] Velayati AA, Masjedi MR, Farnia P, Tabarsi P, Ghanavi J, ZiaZarifi AH, et al. Emergence of new forms of totally drug-resistant tuberculosis bacilli. Chest. 2009;**136**:420-425. DOI: 10.1378/chest.08-2427

[44] Udwadia ZF, Amale RA, Ajbani KK, Rodrigues C. Totally drug-resistant tuberculosis in India. Clinical Infectious Diseases. 2012;**54**(4):579-581. DOI: 10.1093/cid/cir889

[45] Pang X, Cao G, Neuenschwander PF, Haydel SE, Hou G, Howard ST. The β-propeller gene Rv1057 of *Mycobacterium tuberculosis* has a complex promoter directly regulated by both the MprAB and TrcRS two-component systems. Tuberculosis. 2011; **91**:S142-S149. DOI: 10.1016/j.tube.2011.10.024

[46] Kundu M. The role of two-component systems in the physiology of *Mycobacterium tuberculosis*. IUBMB Life. 2018;**70**:710-717. DOI: 10.1002/iub.1872

[47] Neuner EA, Yeh JY, Hall GS, Sekeres J, Endimiani A, Bonomo RA, et al. Treatment and outcomes in carbapenem-resistant *Klebsiella pneumoniae* bloodstream infections. Diagnostic Microbiology and Infectious Disease. 2011;**69**(4):357-362. DOI: 10.1016/j.diagmicrobio.2010.10.013

[48] Zheng J, Lin Z, Sun X, Lin W, Chen Z, Wu Y, et al. Overexpression of OqxAB and MacAB efflux pumps contributes to eravacycline resistance and heteroresistance in clinical isolates of *Klebsiella pneumoniae*. Emerging Microbes & Infections. 2018;7:1-11. DOI: 10.1038/s41426-018-0141-y

[49] Tatsing Foka FE, Ateba CN. Detection of virulence genes in multidrug resistant enterococci isolated from feedlots dairy and beef cattle: Implications for human health and food safety. BioMed Research International. 2019;**2019**:1-13. DOI: 10.1155/2019/5921840

[50] Wardenburg KE, Potter RF, D'Souza AW, Hussain T, Wallace MA, Andleeb S, et al. Phenotypic and genotypic characterization of linezolid-resistant *Enterococcus faecium* from the USA and Pakistan. The Journal of Antimicrobial Chemotherapy. 2019; **74**:3445-3452. DOI: 10.1093/jac/dkz367

[51] Lavilla Lerma L, Benomar N, Valenzuela AS, Mdel CCM, Gálvez A, Abriouel H. Role of EfrAB efflux pump in biocide tolerance and antibiotic resistance of enterococcus faecalis and *Enterococcus faecium* isolated from traditional fermented foods and the effect of EDTA as EfrAB inhibitor. Food Microbiology. 2014;**44**:249-257. DOI: 10.1016/j.fm.2014.06.009

[52] Guzmán Prieto AM, Wijngaarden J, Braat JC, Rogers MRC, Majoor E, Brouwer EC, et al. The two-component

system ChtRS contributes to chlorhexidine tolerance in *Enterococcus faecium*. Antimicrobial Agents and Chemotherapy. 2017;**61**:e02122-16. DOI: 10.1128/AAC.02122-16

[53] Kellogg SL, Little JL, Hoff JS, Kristich CJ. Requirement of the CroRS two-component system for resistance to cell wall-targeting antimicrobials in *Enterococcus faecium*. Antimicrobial Agents and Chemotherapy. 2017;**61**: e02461-16. DOI: 10.1128/AAC.02461-16

[54] Ronald A. The etiology of urinary tract infection: Traditional and emerging pathogens. American Journal of Medicine. 2002;**113**(1):14-19. DOI: 10.1016/S0002-9343(02)01055-0

[55] Kampf G. Adaptive microbial response to low-level benzalkonium chloride exposure. The Journal of Hospital Infection. 2018;**100**:e1-e22. DOI: 10.1016/j.jhin.2018.05.019

[56] He G-X, Zhang C, Crow RR, Thorpe C, Chen H, Kumar S, et al. SugE, a new member of the SMR family of transporters, contributes to antimicrobial resistance in *Enterobacter cloacae*. Antimicrobial Agents and Chemotherapy. 2011;**55**:3954-3957. DOI: 10.1128/AAC.00094-11

[57] Guérin F, Lallement C, Isnard C, Dhalluin A, Cattoir V, Giard J-C. Landscape of resistance-nodulation-cell division (RND)-type efflux pumps in *Enterobacter cloacae* complex. Antimicrobial Agents and Chemotherapy. 2016;**60**:2373-2382. DOI: 10.1128/AAC.02840-15

[58] Annavajhala MK, Gomez-Simmonds A, Uhlemann A-C. Multidrug-resistant *Enterobacter cloacae* complex emerging as a global, diversifying threat. Frontiers in Microbiology. 2019;**10**:44. DOI: 10.3389/fmicb.2019.00044

[59] Coates ARM, Hu Y. Novel approaches to developing new antibiotics for bacterial infections. British Journal of Pharmacology. 2007;**152**:1147-1154. DOI: 10.1038/sj.bjp.0707432

[60] Mujawar S, Mishra R, Pawar S, Gatherer D, Lahiri C. Delineating the plausible molecular vaccine candidates and drug targets of multidrug-resistant *Acinetobacter baumannii*. Frontiers in Cellular and Infection Microbiology. 2019;**9**:203. DOI: 10.3389/fcimb.2019.00203

[61] Pawar S, Ashraf MI, Mujawar S, Mishra R, Lahiri C. In silico identification of the indispensable quorum sensing proteins of multidrug resistant *Proteus mirabilis*. Frontiers in Cellular and Infection Microbiology. 2018;**8**:269. DOI: 10.3389/fcimb.2018.00269

[62] Shrikant P, Chandrajit L. Quorum sensing: An imperative longevity weapon in bacteria. African Journal of Microbiology Research. 2018;**12**:96-104. DOI: 10.5897/AJMR2017.8751

[63] Pawar S, Ashraf I, Mehata KM, Lahiri C. Computational identification of indispensable virulence proteins of *Salmonella* Typhi CT18. In: Curr. Top. Salmonella Salmonellosis. UK: InTech; 2017. DOI: 10.5772/66489

[64] Lahiri C, Pawar S, Sabarinathan R, Ashraf MI, Chand Y, Chakravortty D. Interactome analyses of *Salmonella* pathogenicity islands reveal SicA indispensable for virulence. Journal of Theoretical Biology. 2014;**363**:188-197. DOI: 10.1016/j.jtbi.2014.08.013

[65] Jolly C. Role of the heat shock response and molecular chaperones in oncogenesis and cell death. Journal of the National Cancer Institute. 2000;**92** (19):1564-1572. DOI: 10.1093/jnci/92.19.1564

[66] Takaya A, Tomoyasu T, Matsui H, Yamamoto T. The DnaK/DnaJ chaperone machinery of *Salmonella*

enterica serovar Typhimurium is essential for invasion of epithelial cells and survival within macrophages, leading to systemic infection. Infection and Immunity. 2004;**72**(3):1364-1373. DOI: 10.1128/IAI.72.3.1364-1373.2004

[67] Schröder H, Langer T, Hartl FU, Bukau B. DnaK, DnaJ and GrpE form a cellular chaperone machinery capable of repairing heat-induced protein damage. The EMBO Journal. 1993;**12**(11):4137-4144. DOI: 10.1002/j.1460-2075.1993.tb06097.x

[68] Zylicz M, Wawrzynow A. Insights into the function of Hsp70 chaperones. IUBMB Life. 2001;**51**(5):283-287. DOI: 10.1080/152165401317190770

[69] Mayer MP, Rudiger S, Bukau B. Molecular basis for interactions of the DnaK chaperone with substrates. Biological Chemistry. 2000;**381**(9-10):877-885. DOI: 10.1515/BC.2000.109

[70] Chiappori F, Fumian M, Milanesi L, Merelli I. DnaK as antibiotic target: Hot spot residues analysis for differential inhibition of the bacterial protein in comparison with the human Hsp70. PLoS One. 2015;**10**(4):e0124563. DOI: 10.1371/journal.pone.0124563

[71] Singh VK, Utaida S, Jackson LS, Jayaswal RK, Wilkinson BJ, Chamberlain NR. Role for dnaK locus in tolerance of multiple stresses in *Staphylococcus aureus*. Microbiology. 2007;**153**:3162-3173. DOI: 10.1099/mic.0.2007/009506-0

[72] Yamaguchi Y, Tomoyasu T, Takaya A, Morioka M, Yamamoto T. Effects of disruption of heat shock genes on susceptibility of *Escherichia coli* to fluoroquinolones. BMC Microbiology. 2003;**3**:16. DOI: 10.1186/1471-2180-3-16

[73] Sun J, Deng Z, Yan A. Bacterial multidrug efflux pumps: Mechanisms, physiology and pharmacological exploitations. Biochemical and Biophysical Research Communications. 2014;**453**:254-267. DOI: 10.1016/j.bbrc.2014.05.090

[74] Piddock LJV. Multidrug-resistance efflux pumps? Not just for resistance. Nature Reviews. Microbiology. 2006;**4**:629-636. DOI: 10.1038/nrmicro1464

[75] Poole K. Efflux pumps as antimicrobial resistance mechanisms. Annals of Medicine. 2007;**39**:162-176. DOI: 10.1080/07853890701195262

[76] Nishino K, Nikaido E, Yamaguchi A. Regulation and physiological function of multidrug efflux pumps in *Escherichia coli* and *Salmonella*. Biochimica et Biophysica Acta, Proteins and Proteomics. 2009;**1794**(5):834-843. DOI: 10.1016/j.bbapap.2009.02.002

[77] Masuda N, Sakagawa E, Ohya S, Gotoh N, Tsujimoto H, Nishino T. Substrate specificities of MexAB-OprM, MexCD-OprJ, and MexXY-OprM efflux pumps in *Pseudomonas aeruginosa*. Antimicrobial Agents and Chemotherapy. 2000;**44**:3322-3327. DOI: 10.1128/AAC.44.12.3322-3327.2000

[78] Capra EJ, Laub MT. Evolution of two-component signal transduction systems. Annual Review of Microbiology. 2012;**66**:325-347. DOI: 10.1146/annurev-micro-092611-150039

[79] Mizuno T. Compilation of all genes encoding two-component phosphotransfer signal transducers in the genome of *Escherichia coli*. DNA Research. 1997;**4**(2):161-168. DOI: 10.1093/dnares/4.2.161

[80] Hirakawa H, Nishino K, Hirata T, Yamaguchi A. Comprehensive studies of drug resistance mediated by overexpression of response regulators of two-component signal transduction systems in *Escherichia coli*. Journal of Bacteriology. 2003;**185**(6):1851-1856. DOI: 10.1128/JB.185.6.1851-1856.2003

[81] Hirakawa H, Inazumi Y, Masaki T, Hirata T, Yamaguchi A. Indole induces the expression of multidrug exporter genes in *Escherichia coli*. Molecular Microbiology. 2004;**55**:1113-1126. DOI: 10.1111/j.1365-2958.2004.04449.x

[82] Nishino K, Yamaguchi A. Overexpression of the response regulator evgA of the two-component signal transduction system modulates multidrug resistance conferred by multidrug resistance transporters. Journal of Bacteriology. 2001;**83**(4): 1455-1458. DOI: 10.1128/JB.183.4.1455-1458.2001

[83] Nishino K, Yamaguchi A. EvgA of the two-component signal transduction system modulates production of the YhiUV multidrug transporter in *Escherichia coli*. Journal of Bacteriology. 2002;**184**(8):2319-2323. DOI: 10.1128/JB.184.8.2319-2323.2002

[84] Zhang Y, Xiao M, Horiyama T, Zhang Y, Li X, Nishino K, et al. The multidrug efflux pump MdtEF protects against nitrosative damage during the anaerobic respiration in *Escherichia coli*. Journal of Biological Chemistry. 2011; **286**(30):26576-26584. DOI: 10.1074/jbc.M111.243261

[85] Deng Z, Shan Y, Pan Q, Gao X, Yan A. Anaerobic expression of the gadE-mdtEF multidrug efflux operon is primarily regulated by the two-component system ArcBA through antagonizing the H-NS mediated repression. Frontiers in Microbiology. 2013;**4**:194. DOI: 10.3389/fmicb.2013.00194

[86] Nishino K, Nikaido E, Yamaguchi A. Regulation of multidrug efflux systems involved in multidrug and metal resistance of *Salmonella enterica* serovar Typhimurium. Journal of Bacteriology. 2007;**189**(24):9066-9075. DOI: 10.1128/JB.01045-07

[87] García Véscovi E, Soncini FC, Groisman EA. Mg^{2+} as an extracellular signal: Environmental regulation of *Salmonella* virulence. Cell. 1996;**84**: 165-174. DOI: 10.1016/S0092-8674(00)81003-X

[88] Bearson BL, Wilson L, Foster JW. A low pH-inducible, PhoPQ-dependent acid tolerance response protects *Salmonella* Typhimurium against inorganic acid stress. Journal of Bacteriology. 1998;**180**(9):2409-2417

[89] Sivakumar D, Lahiri C, Chakravortty D. Computational studies on histidine kinase protein BaeS to target multidrug-resistant *Salmonella*. Medicinal Chemistry Research. 2013;**22**:1804-1811. DOI: 10.1007/s00044-012-0188-6

[90] Tiwari S, Jamal SB, Hassan SS, Carvalho PVSD, Almeida S, Barh D, et al. Two-component signal transduction systems of pathogenic bacteria as targets for antimicrobial therapy: An overview. Frontiers in Microbiology. 2017;**8**:1878. DOI: 10.3389/fmicb.2017.01878

[91] Marchand I, Damier-Piolle L, Courvalin P, Lambert T. Expression of the RND-type efflux pump AdeABC in *Acinetobacter baumannii* is regulated by the AdeRS two-component system. Antimicrobial Agents and Chemotherapy. 2004;**48**(9):3298-3304. DOI: 10.1128/AAC.48.9.3298-3304.2004

[92] Li XZ, Zhang L, Poole K. SmeC, an outer membrane multidrug efflux protein of *Stenotrophomonas maltophilia*. Antimicrobial Agents and Chemotherapy. 2002;**46**(2):333-343. DOI: 10.1128/AAC.46.2.333-343.2002

[93] Bem AE, Velikova N, Pellicer MT, Van Baarlen P, Marina A, Wells JM. Bacterial histidine kinases as novel antibacterial drug targets. ACS Chemical Biology. 2015;**10**(1):213-224. DOI: 10.1021/cb5007135

[94] Ikuma K, Decho AW, Lau BLT. The extracellular bastions of Bacteria—A biofilm way of life. Nature Education Knowledge. 2013;**4**(2):2

[95] Wojtyczka RD, Orlewska K, Kepa M, Idzik D, Dziedzic A, Mularz T, et al. Biofilm formation and antimicrobial susceptibility of *Staphylococcus epidermidis* strains from a hospital environment. International Journal of Environmental Research and Public Health. 2014;**11**(5):4619-4633. DOI: 10.3390/ijerph110504619

[96] Davey ME, O'toole GA. Microbial biofilms: From ecology to molecular genetics. Microbiology and Molecular Biology Reviews. 2000;**64**:847-867. DOI: 10.1128/MMBR.64.4.847-867.2000

[97] Sutherland IW. Biofilm exopolysaccharides: A strong and sticky framework. Microbiology. 2001;**147**:3-9. DOI: 10.1099/00221287-147-1-3

[98] Staley C, Dunny GM, Sadowsky MJ. Environmental and animal-associated enterococci. Advances in Applied Microbiology. 2014;**87**:147-186. DOI: 10.1016/B978-0-12-800261-2.00004-9

[99] May TB, Shinabarger D, Maharaj R, Kato J, Chu L, Devault JD, et al. Alginate synthesis by *Pseudomonas aeruginosa*: A key pathogenic factor in chronic pulmonary infections of cystic fibrosis patients. Clinical Microbiology Reviews. 1991;**4**(2): 191-206. DOI: 10.1128/CMR.4.2.191

[100] Costerton JW, Lewandowski Z, Caldwell DE, Korber DR, Lappin-Scott HM. Microbial biofilms. Annual Review of Microbiology. 1995;**49**:711-745. DOI: 10.1146/annurev.mi.49.100195.003431

[101] Danese PN, Pratt LA, Kolter R. Exopolysaccharide production is required for development of *Escherichia coli* K-12 biofilm architecture. Journal of Bacteriology. 2000;**182**:3593-3596. DOI: 10.1128/JB.182.12.3593-3596.2000

[102] Lewis T, Loman NJ, Bingle L, Jumaa P, Weinstock GM, Mortiboy D, et al. High-throughput whole-genome sequencing to dissect the epidemiology of *Acinetobacter baumannii* isolates from a hospital outbreak. Journal of Hospital Infection. 2010;**75**(1):37-41. DOI: 10.1016/j.jhin.2010.01.012

[103] Fajardo A, Martínez-Martín N, Mercadillo M, Galán JC, Ghysels B, Matthijs S, et al. The neglected intrinsic resistome of bacterial pathogens. PLoS One. 2008;**3**(2):e1619. DOI: 10.1371/journal.pone.0001619

[104] Stickler DJ, King JB, Winters C, Morris SL. Blockage of urethral catheters by bacterial biofilms. The Journal of Infection. 1993;**27**:133-135. DOI: 10.1016/0163-4453(93)94620-Q

[105] Blanchette KA, Wenke JC. Current therapies in treatment and prevention of fracture wound biofilms: Why a multifaceted approach is essential for resolving persistent infections. Journal of Bone and Joint Infection. 2018;**3**(2): 50-67. DOI: 10.7150/jbji.23423

[106] Nickel JC, Ruseska I, Wright JB, Costerton JW. Tobramycin resistance of *Pseudomonas aeruginosa* cells growing as a biofilm on urinary catheter material. Antimicrobial Agents and Chemotherapy. 1985;**27**(4):619-624. DOI: 10.1128/AAC.27.4.619

[107] Nickel JC, Downey JA, Costerton JW. Ultrastructural study of microbiologic colonization of urinary catheters. Urology. 1989;**34**:284-291. DOI: 10.1016/0090-4295(89)90327-0

[108] Gristina AG, Dobbins JJ, Giammara B, Lewis JC, Devries WC. Biomaterial-centered sepsis and the total artificial heart: Microbial adhesion vs tissue integration. JAMA: The Journal of the American Medical Association. 1988;**259**(6):870-874. DOI: 10.1001/jama.1988.03720060038027

[109] Akyıldız İ, Take G, Uygur K, Kızıl Y, Aydil U. Bacterial biofilm formation in the middle-ear mucosa of chronic otitis media patients. Indian Journal of Otolaryngology and Head and Neck Surgery. 2013;**65**(Suppl 3): 557-561. DOI: 10.1007/s12070-012-0513-x

Chapter 2

Mechanisms of Resistance to Quinolones

Sandra Georgina Solano-Gálvez,
María Fernanda Valencia-Segrove,
María José Ostos Prado, Ana Berenice López Boucieguez,
Diego Abelardo Álvarez-Hernández
and Rosalino Vázquez-López

Abstract

Antimicrobial resistance is a worldwide problem. Various pathogenic bacteria can be resistant to one or several antibiotics, resulting in a serious public health problem. Isolation of pathogenic bacteria resistant to multiple last-generation antibiotics from hospital samples have been reported. In that sense, the isolation of pathogenic strains resistant to members of the quinolone family, from clinical samples, is an increasing phenomenon. Quinolones are a group of synthetic broad-spectrum antimicrobials, whose mechanism of action is the inhibition of DNA gyrase and topoisomerase IV, with the consequent DNA breakdown and cell death due to genotoxic damage. Three mechanisms have been determined by which bacteria can be resistant to quinolones: (1) Chromosomal mutations in coding genes (mutations that alter the objectives of the drug). (2) Mutations associated with the reduction of the intracytoplasmic concentration of quinolones. (3) Plasmid-mediated quinolone resistance genes (plasmids that protect cells from the lethal effects of quinolones). In this chapter, we analyze each of them and provide the most current connections and investigations of these processes.

Keywords: antibiotic resistance, quinolones, fluoroquinolones, DNA topoisomerase IV, genotoxic damage

1. Background

Antimicrobial resistance has become a serious public health problem in recent years. This problem has been increasing and is currently a truly global crisis that offers one of the worst forecasts of catastrophic scenarios in public health worldwide.

A sign of the seriousness of the problem is the fact that World Health Organization (WHO)'s new Global Antimicrobial Surveillance System (GLASS) reported the widespread occurrence of antibiotic resistance among 500,000 people with suspected bacterial infections across 22 countries [1].

Likewise, Centers for Disease Control (CDC)'s Antibiotic Resistance Threats in the United States (US), in 2019 (2019 AR Threats Report), reported that more than

2.8 million antibiotic-resistant infections occur in the US each year, and more than 35,000 people die as a result. Besides, 223,900 cases of *Clostridium difficile* occurred in 2017 and at least 12,800 people died [2].

Many bacteria produce important infections in human health, either due to community-acquired infections, nosocomial infections, or at intensive care units. Among these, many have an important phenotypic profile of antibiotic resistance. For example, *Staphylococcus aureus, Enterococcus* spp., *Enterobacteriaceae* (other than *Salmonella* and *Shigella*), *Pseudomonas aeruginosa,* and *Acinetobacter* spp. [3, 4].

To classify these microorganisms according to the degree of resistance and acquired resistance profiles, a group of experts in the field of antimicrobial resistance in joint work with the European Center for the Prevention of Diseases and Control (ECDC) and the CDC established the definitions and characteristics among resistant bacteria: multidrug-resistant (MDR), extensively drug-resistant (XDR), and pan drug-resistant (PDR) bacteria [3, 4].

To establish objective parameters of the phenotypic resistance profile in each of these bacteria, epidemiologically significant antimicrobial categories were established. These categories were established based on the documents and cut-off points of the Clinical Laboratory Standards Institute (CLSI), the European Antimicrobial Sensitivity Testing Committee (EUCAST), and the US Food and Drug Administration (FDA) [3, 4].

Based on the new limits and definitions: MDR bacteria possess acquired resistance to at least one antibiotic of three or more categories; XDR bacteria possess resistance to at least one antibiotic of almost all categories, except one or two of them; and PDR bacteria are resistant to all agents of all categories of antimicrobials [3, 4].

Antimicrobial resistance has been observed in all families of antibiotics, including the latest generation and intrahospital antibiotics such as quinolones.

The wide use of quinolones in clinical practice includes the administration of the antibiotic in prophylaxis, in neutropenic patients with cancers, in cirrhotic patients at risk for spontaneous bacterial peritonitis, and in urologic surgery, among others. In many of these cases, strains with varying degrees of resistance to quinolones have been isolated [5, 6].

2. History

In 1962, quinolones were discovered as an important treatment for various pathological manifestations. The first one was nalidixic acid, which was synthetically produced by George Lesher at the Sterling-Winthrop Research Institute. It was synthesized from the isolation of chloro-1-ethyl-1,4-dihydro-4-oxo-3-quinoline carboxylic acid years before, as a product derived from the synthesis of chloroquine [7]. Its origin dates back to the use of chloroquine as an antimalarial agent. It was until years after its development that nalidixic acid was approved for the treatment of urinary tract infections by Gram-negative bacteria. This compound does not have an important effect on Gram-positive bacteria, in addition to having a certain cytotoxic effect on the gastrointestinal tract and the central nervous system. Its effect on Gram-negative bacteria is characteristic of the first generation of quinolones [8].

3. Epidemiology

The indiscriminate prescription of quinolones worldwide has led to a rapid increase in bacterial resistance. *Acinetobacter* spp., *Campylobacter* spp., *Capnocytophaga* spp., *Clostridium* spp., *Escherichia coli, Klebsiella pneumoniae, Mycobacterium tuberculosis,*

Bacteria	Mean resistance percentage		Country with the lowest resistance percentage		Country with the highest resistance percentage	
	2015	2018	2015	2018	2015	2018
Acinetobacter spp.	43.7%	36.2%	Belgium (0%)	Norway (0%)	Greece (94.9%)	Croatia (96.1%)
Escherichia coli	22.8%	25.3%	Iceland (6.8%)	Finland (11.4%)	Cyprus (45.5%)	Cyprus (42.4%)
Klebsiella pneumoniae	29.7%	31.6%	Iceland (2.9%)	Iceland (0%)	Slovakia (70%)	Poland (68.2%)
Pseudomonas aeruginosa	19.3%	19.7%	Estonia (0%)	Malta (0%)	Romania (59%)	Slovakia (52.4%)

The EARS-Net report does not contain information about quinolone resistance to other bacteria. Adapted from: EARS-Net 2015 and Ecdc. SURVEILLANCE REPORT. 2018 [11, 12].

Table 1.
Profile of resistance to quinolones of European countries (2015 vs. 2018).

Neisseria gonorrhea, Proteus mirabilis, P. aeruginosa, Salmonella spp., *S. aureus,* and *Streptococcus pneumoniae,* among others, have been reported as resistant [7, 9, 10].

The ECDC collects and reports through the European Antimicrobial Resistance Surveillance Network (EARS-Net) information of seven bacterial pathogens that commonly cause infections in humans: *Acinetobacter* spp., *Enterococcus faecalis, Enterococcus faecium, Escherichia coli, Klebsiella pneumoniae, Staphylococcus aureus, Streptococcus pneumoniae,* and *Pseudomonas aeruginosa.* Information comparing their profile of resistance to quinolones in Europe between 2015 and 2018 can be found in **Table 1** [11, 12].

4. The structure of quinolones

The structure of quinolones derives from two types of rings, a naphthyridine core with a nitrogen molecule in positions 1 and 8. Through this structure, the compound is limited to being used as a therapy against Gram-negative bacteria. However, it has been shown that by inserting a cyclopropyl group in the first position of the nitrogen ring, an effect is achieved not only on Gram-negative bacteria but also on Gram-positive ones (**Figure 1**) [13, 14].

The second generation was developed in 1980, from the addition of a fluorine atom at position six, resulting in fluoroquinolones. These have higher activity in Gram-negative bacteria, as well as Gram-positive bacteria. Some fluoroquinolones can inhibit all Gram-negative organisms. Quinolones with piperazine on carbon 7 are effective in Gram-negative bacteria and the signaling of topoisomerase 4 (**Figure 1**) [13, 15–17].

Later, the third generation arises by adding certain molecules in the rings, such as the cyclopropyl ring in the first position of nitrogen, improving the activity in Gram-positive bacteria. Some of these modifications achieved sensitivity in organism resistant to different antibiotics, including *Streptococcus pneumoniae*. Other benefits of this generation are a longer life in serum and activity against anaerobic organisms (**Figure 1**) [7, 18, 19].

The fourth generation was later developed by incorporating nitrogen in the eighth position, resulting in a broad-spectrum antibiotic. Its action in some Gram-positive organisms is more effective compared to the other generations; however, its activity in anaerobic organisms is limited. It has a superior bacterial selectivity to avoid a high level of resistance and its toxic effects are less unfavorable than in the other generations [7, 8]. Thanks to the modifications made to the quinolones,

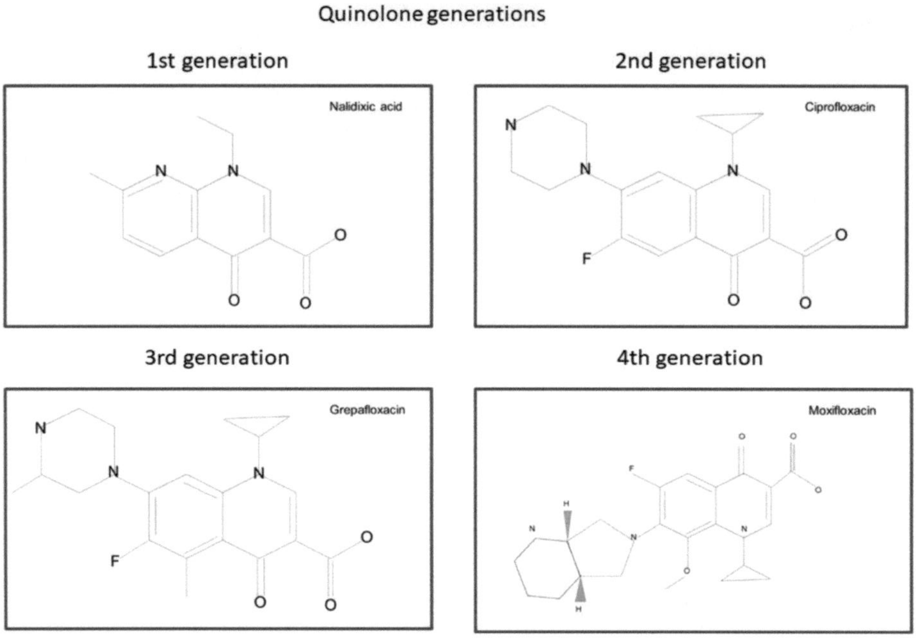

Figure 1.
Molecular structure of representative members of each quinolone generation. Based on PubChem public archive https://www.ncbi.nlm.nih.gov/pcsubstance [14, 17, 19, 20].

Generation		Compounds	Activity spectrum
1		Nalidixic acid	Gram-negative bacteria (not *Pseudomonas* spp.)
2	2a	Ciprofloxacin, enoxacin, norfloxacin	Gram-negative bacteria and atypical pathogens (*Mycoplasma pneumoniae* and *Chlamydia pneumoniae*)
	2b	Levofloxacin, lomefloxacin, ofloxacin	Gram-negative bacteria, Gram-positive bacteria (not *Streptococcus pneumoniae*), and atypical pathogens
3		Clinafloxacin, gatifloxacin, grepafloxacin, sparfloxacin.	Gram-negative bacteria, Gram-positive bacteria (*Streptococcus pneumoniae*) and improved activity against atypical pathogens
4		Gatifloxacin, gemifloxacin, moxifloxacin, trovafloxacinin	Gram-negative bacteria, Gram-positive bacteria (*Streptococcus pneumoniae*) and improved activity against atypical and anaerobic pathogens

Adapted from: Pham TDM, Ziora ZM, Blaskovich MAT [7].

Table 2.
Classification of quinolones.

an improvement in its pharmacokinetics and pharmacodynamics has been obtained, thus optimizing absorption, metabolism, and elimination, achieving lower toxicity and superiority in the mechanisms of action. It has also been possible to modify the half-life of the drug making only one dose per day necessary (**Figure 1**) [20].

Currently, nine fluoroquinolones have been approved in the US while others continue to be used in clinical trials. Information regarding the generations, compounds, and spectrum of activity can be found in **Table 2**.

It has been reported that several agents such as *Escherichia coli, Klebsiella pneumoniae, Neisseria gonorrhoeae,* or *Staphylococcus aureus* have presented significant resistance to quinolones [21]. Plasmid-mediated quinolone resistance (PMQR) was a completely unexpected event since it was thought that the only mutation would occur in genes encoding topoisomerase II identification. Currently, the resistance mechanism is multifactorial. However, the most common quinolone resistance mechanism is topoisomerase mutations [15, 22].

The excessive use of this type of drug has caused the incidence rates of hypersensitivity to increase more and more, taking the second place of antibiotics with a greater number of hypersensitivity reactions in in-hospital patients. The main agents that cause hypersensitivity are ciprofloxacin, levofloxacin, and moxifloxacin. This has positioned quinolones as the non-beta-lactam antibiotics with the highest incidence of hypersensitivity reactions [23, 24].

5. Mechanism of action

The mechanism of action of quinolones is based on the inhibition of bacterial topoisomerases II and IV. Topoisomerases are enzymes responsible for maintaining the tertiary structure of DNA during various cellular processes, such as synthesis, replication, condensation, and decondensation of DNA, among others [25–29].

Topoisomerase II, also known as DNA gyrase, is considered a negative supercoiling enzyme, which means that it cuts the two strands of DNA and propitiates that the DNA is twisted to the left producing a twist in a way contrary to the direction of the double helix. This enzyme participates in the DNA winding and relaxation during various processes, mainly in the synthesis and replication of DNA [30, 31].

The DNA gyrase consists of a heterotetramer, which is formed by two GyrA subunits and two GyrB subunits. The GyrA subunits participate in the union with the DNA and are responsible for making the double helix cuts. The GyrB subunits possess ATPase activity [30].

Topoisomerase IV is responsible for preventing the chromatids from being chained, meaning it participates in the separation of daughter chromatids after DNA replication [32].

Like DNA gyrase, topoisomerase IV is made up of a tetramer. It has two ParC subunits and two ParE subunits. These subunits possess homologous activity of GyrA and GyrB, respectively [32].

When quinolones interact and inhibit topoisomerase II and IV, it induces DNA breakdown and cell death due to genotoxic damage [27–29].

6. Resistance mechanisms

To counteract the effect of quinolones, bacteria have developed various resistance mechanisms to these antibiotics. Bacterial resistance to quinolones is mainly based on three points (**Table 3**, **Figure 2**):

1. Chromosomal mutations in coding genes (mutations that alter the objectives of the drug).

2. Mutations associated with the reduction of the intracytoplasmic concentration of quinolones.

3. PMQR genes (plasmids that protect cells from the lethal effects of quinolones) [33].

Mechanism	Description		
Chromosomal mutations in coding genes	Occurs due to errors in the replication of the genes encoding the GyrA subunits of DNA gyrase and ParC of topoisomerase IV		
Mutations associated with the reduction of the intracytoplasmic concentration of quinolones	Occurs due to mutations that lead to a decrease in the intracytoplasmic concentration of the antibiotic. It may happen through:		
	Overexpression of efflux pumps from the resistance-nodulation-cell division	Both	Reduction of the membrane permeability by downregulation of extra-membrane proteins
Plasmid-mediated quinolone resistance genes	Occurs due to the activation of plasmid-mediated quinolone resistance genes. Among them are:		
	Qnr's encode proteins that protect DNA gyrase and topoisomerase IV	AAC(6′)-lb-cr acetylates quinolones with an appropriate amino nitrogen target	QepA and OqxAB, which increase the outflow of quinolones through efflux pumps

Adapted from: Álvarez-Hernández DA, Garza-Mayén GS, Vázquez-López R. Quinolones. Nowadays perspectives and mechanisms of resistance [34].

Table 3.
Mechanisms of resistance to quinolones.

6.1 Chromosomal mutations in coding genes (mutations that alter the objectives of the drug)

The quinolone resistance associated with chromosomal mutations occurs due to errors in the replication of the genes encoding the GyrA subunits of DNA gyrase and ParC of topoisomerase IV [33, 35].

In the amino acid sequences of the GyrA and ParC subunits, there are specific regions that interact with the DNA. In these regions, there are conserved domains called quinolone resistance determining region (QRDR) [31, 35–39].

It is precisely in the sequences that code for each of the QRDR domains of the GyrA and ParC subunit genes, where such mutations occur [31, 35–39].

It has been reported that quinolone resistance may also occur due to mutations in the genes encoding the GyrB and ParE subunits; however, they do not occur so frequently and their clinical value appears to be very limited [35, 40, 41].

There is evidence that in Gram-negative bacteria, DNA gyrase turns out to be more susceptible to inhibition than topoisomerase IV. On the other hand, in Gram-positive bacteria, the opposite phenomenon occurs; that is, that topoisomerase IV is more susceptible to inhibition than gyrase. However, certain bacteria show the opposite effect, being the exception to the rule [31, 42, 43].

Therefore, we can affirm that the phenomenon of resistance in the majority of Gram-negative bacteria occurs mainly in GyrA, while in most Gram-positive bacteria the inhibition of ParC is the most important [31, 42, 43].

Summarizing, mutations that occur in the sequences encoding the QRDR domains in both GyrA-ParC and GyrB-ParE favor a decrease in the binding affinity of quinolones with the DNA–DNA gyrase and DNA-topoisomerase IV complex [33, 35].

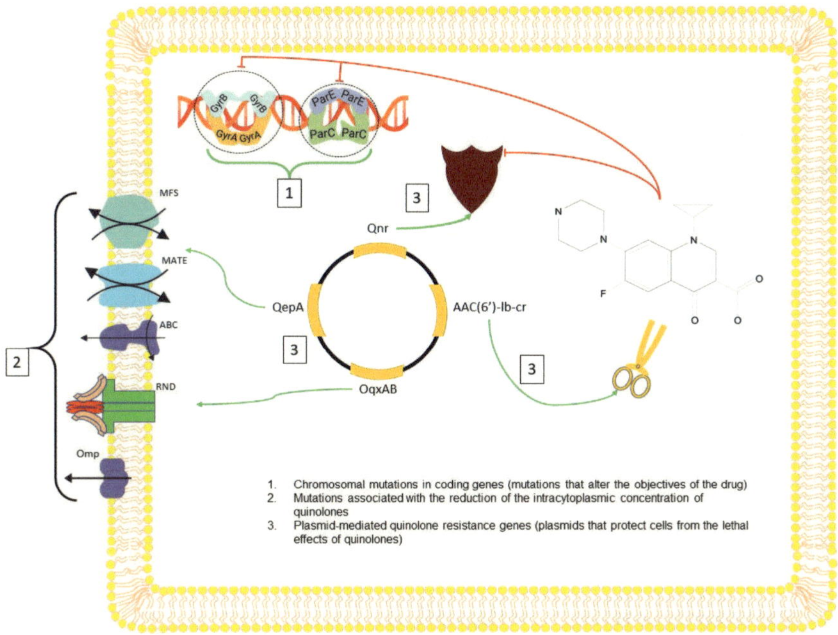

Figure 2.
Schematic representation of the mechanisms of bacterial resistance to quinolones. Based on Susana Correia et al. [22].

6.2 Mutations associated with the reduction of the intracytoplasmic concentration of quinolones

Another important quinolone resistance mechanism consists in the ability of the bacteria to decrease the intracytoplasmic concentration of the antibiotic; this decrease in concentration is determined by certain mutations.

This phenomenon is achieved through three mechanisms:

1. Efflux pumps that promote the active transport of quinolones to the outside of the bacterial cell.

2. Decreased membrane permeability toward the antibiotic.

3. A combination of both mechanisms.

It has been described that only efflux pumps participate in Gram-positive bacteria as mechanisms to reduce the intracytoplasmic concentration of quinolones since there is no evidence that the decrease in cytoplasmic membrane permeability participates in this type of bacteria [44].

On the other hand, Gram-negative bacteria do have both mechanisms and participate in a complementary way with one another, the decrease in permeability in the cytoplasmic membrane being the most important for these bacteria [45].

These two mechanisms involved in the decrease of the intracytoplasmic concentration of quinolones are not induced by the drugs themselves. There is evidence that these two mechanisms occur because of mutations in genes that encode regulatory proteins that control transcription of the outflow pump or genes that code for porin synthesis [35, 46].

6.2.1 Mutations associated with the reduction of the intracytoplasmic concentration of quinolones in Gram-positive bacteria

This resistance mechanism in Gram-positive bacteria is associated with the presence of chromosomally encoded efflux pumps that decrease the intracytoplasmic concentration of the antibiotic, giving the bacteria the characteristic of being MDR.

Efflux pumps are classified into two groups: primary active transporters and secondary active transporters [47].

The primary active transporter proteins are pumps that use ATP as a source of energy. This type of primary active transporter integrates the members of the ATP-binding cassette (ABC) superfamily [48–50].

On the other hand, the secondary active transporter proteins use the energy obtained by the difference of chemical gradients formed by either protons or ions, for example, sodium ions [48, 49].

Four types of secondary active transporter proteins have been identified: [47–49].

1. The small multidrug-resistance (SMR) family

2. The major facilitator superfamily (MFS)

3. Multidrug and toxic compound extrusion (MATE) family

4. The resistance-nodulation-cell division (RND) superfamily.

6.2.1.1 SMR (the small multidrug-resistance family)

Members of this family are proteins made up of an antiparallel dimer. Each monomer of this dimer has four transmembrane helices (TM1, TM2, TM3, and TM4). The TM 1 to M3 helices comprise the substrate binding pocket, while each TM4 helix is responsible for SMR TM4-TM4 dimerization [51–53].

The members of the SMR family are associated with resistance to various toxic compounds and some antibiotics; however, they do not appear to play a relevant role in resistance to quinolones.

6.2.1.2 MFS (the major facilitator superfamily)

Concerning efflux pumps related to the intracytoplasmic decrease in quinolone and consequently linked to resistance to this drug, they are efflux pumps that are part of the MFS. Three members of this family associated with quinolone resistance have been identified: NorA, NorB [50], and NorC [54]. Overexpression of each of three efflux pumps increases resistance to quinolones four to eight times [33].

6.2.1.2.1 NorA

The chromosomal gene that codes for NorA could be identified in 1986 from the isolation of *Staphylococcus aureus* obtained from a urine sample from a patient who had received treatment with norfloxacin at Teikyo University Hospital Japan [55]. It has been observed that NorA participates in the pumping of various quinolones, mainly ciprofloxacin and norfloxacin [56, 57].

Subsequent studies of genetic diversity described three alleles for the *NorA* gene [58]: *NorAI* (Yoshida), *NorAII* (Noguchi), and *NorAIII* (Kaatz). A correlation has been observed between the different types of NorA alleles and specific lineages of *S. aureus*. This fact suggests that there is a correlation between the NorA variants and the population structure (lineages) of this bacterium [58].

6.2.1.2.2 NorB

It has been described that the expression of the efflux pump NorB gives certain bacteria (e.g., *Staphylococcus aureus*) the adaptability in tissue infection conditions, even in the absence of antibiotics. This fact occurs because NorB gives *Staphylococcus aureus* the ability to eliminate antibacterial substances present in the abscess and produced as a defense mechanism by the host. In this way, NorB not only participates in the quinolone resistance mechanism but also contributes to the pathophysiology of certain infections [59].

6.2.1.2.3 NorC

The efflux pump *Norc* enhances the exit of quinolones such as ciprofloxacin, garenoxacin moxifloxacin, and sparfloxacin out of the bacterial cell. Its expression is regulated negatively by MgrA [54].

Many regulatory proteins participate in a complex regulatory process in the gene expression of NorA, NorB, and NorC. One of these regulatory proteins is MgrA, which shows the ability to bind to the NorA promoter region. The overexpression of MgrA causes the inhibition of the expression of NorA, NorB, and NorC, in the opposite, resistance to quinolones is associated with a low activity of *MgrA* and the consequent overproduction of NorA, NorB, and NorC that will promote a decrease in the intracytoplasmic concentration of the drug [54, 60–62].

There is evidence that MgrA activity could be determined by environmental conditions in which the bacterium is found. Acid conditions, oxidative, as well as the presence of iron, could alter the activity of MgrA and consequently the expression of NorA, NorB, and NorC and its effect on the pumping of quinolone and its concentration in the bacterial cytoplasm [35, 59, 63–65].

On the other hand, another transcriptional regulator, called NorG, which activates the expression of NorA and NorB but suppresses the expression of NorC, has been described. It is important to understand that the regulation of the gene expression of *NorA*, *NorB*, and *NorC* results from a complex molecular framework where both activators and inhibitors participate and the balance between them, as well as the environmental and nutritional conditions in which the bacteria develops, will give as a result the resistance or the lack of it to quinolones [35, 61, 62, 66].

6.2.1.3 Other members of the MFS (major facilitator superfamily)

6.2.1.3.1 MdeA

MdeA gen was identified in an open reading frame (ORF) expression library of the *S. aureus* genome. The efflux pump protein MdeA belongs to the MFS using the proton motive force to energize the transport of its substrates [67, 68].

MdeA confers resistance to the biocides benzalkonium chloride, dequalinium, tetraphenylphosphonium, and to the dye ethidium bromide [67]. MdeA also confers resistance to multiple antibiotics among which are fusidic acid, mupirocin, novobiocin, and virginiamycin, and to some extent toward ciprofloxacin and norfloxacin [67, 68].

6.2.1.3.2 SdrM

In 2006 Yamada et al. cloned a new gene called *SA1972* isolated from *Staphylococcus aureus*. The product obtained was called SdrM and it was proven that it conferred resistance to the bacteria against, acriflavine, ethidium bromide, and norfloxacin. SdrM was classified as an efflux pump belonging to the MFS [69].

6.2.1.3.3 QacB (III)

The *qacA* and *qacB* genes that code for efflux pump proteins (QacA and QacB, respectively) are present in methicillin-resistant *Staphylococcus aureus* (MRSA). The efflux pump QacA has two isoforms, while the pump QacB has four known as QacBI, QacBII, QacBIII, and QacBIV. It has been observed that the QacBIII variant confers resistance to *S. aureus* to fluoroquinolones [70].

6.2.1.4 MATE (multidrug and toxic compound extrusion family)

6.2.1.4.1 MepA

The efflux pump MepA belongs to the multidrug and toxic compound extrusion (MATE) family. MepA gives the bacterium a phenotypic MDR profile associated with low-level resistance to some quaternary ammonium compounds. It also confers resistance to certain antibiotics, mainly toward glycylcyclines and to a lesser extent resistance to ciprofloxacin and norfloxacin [71–73].

In addition to the efflux pump described above, there are other transporters in Gram-positive bacteria that participate in the decrease in the intracytoplasmic concentration of quinolones in the bacterial cell, participating in resistance to this drug. Some of these transporters are LmrS, Bmr, Bmr3 and Blt, PmrA66, LmrP67, PatAB69, SatAB70, LmrA71, FepA, FepR, and TetR [35].

6.2.2 Mutations associated with the reduction of concentration in Gram-negative bacteria

6.2.2.1 RND (resistance-nodulation-cell division superfamily)

Gram-negative bacteria use efflux pumps belonging to the RND superfamily as the main mechanism of resistance to quinolones. The efflux pump RND pumps are a molecular complex consisting of three elements (**Figure 3**) [49, 74–77]:

1. In the inner membrane is RND pump protein.

2. An adapter protein from the MFP (membrane fusion protein) family located in the periplasmic space.

3. In the outer membrane is an outer membrane channel protein (OMP) belonging to the outer membrane factor (OMF) family.

The adapter protein MFP links the pump RND and the OMF protein [49, 74–77]. In *E. coli*, the presence of five RND efflux transporters has been reported:

1. AcrAB [78, 79]

2. AcrAD [80, 81]

3. AcrEF [82]

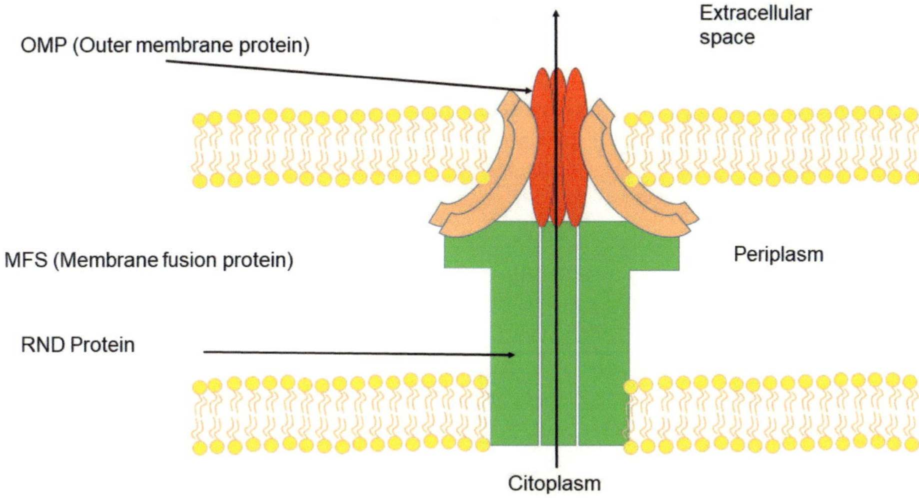

Figure 3.
Schematic representation of molecular structure of RND (resistance-nodulation-cell division superfamily). Based on Eun-Hae Kim et al. [77].

4. MdtABC [83, 84]

5. MdtEF [85, 86]

6.2.2.2 AcrAB-TolC [acriflavine (Acr) efflux system]

The AcrAB-TolC or acriflavine (Acr) efflux system consists of three elements [75, 87]:

1. The outer-membrane channel TolC

2. In the periplasmic space is the AcrA protein, which bridges these two integral membrane proteins

3. In the inner membrane is the secondary transporter AcrB.

There is evidence that the ratio between the proteins that make up this complex is 3: 6: 3, comprising an AcrB trimer, an AcrA hexamer, and a TolC trimer [75, 87].

It has been shown that various dyes can be accommodated in the transmembrane domain of the Acr efflux system, as well as doxorubicin, minocycline, and quinolone molecules [88, 89].

6.2.2.2.1 AcrAD

AcrAD is an antibiotic efflux pump complex of the RND type. It provides resistance to aminoglycosides such as amikacin, gentamicin, and tobramycin. There is no known effect on quinolone resistance [80, 90].

6.2.2.2.2 AcrEF

AcrEF is an antibiotic efflux pump complex of the resistance-nodulation-cell division (RND) type. It provides resistance to cephalosporins, cephamycins, fluoroquinolones, and penams [91, 92].

6.2.2.2.3 MdtABC

MdtABC is an antibiotic efflux pump complex of the resistance-nodulation-cell division (RND) type. It provides resistance to aminocoumarins, which have a mechanism of action similar to quinolones [93, 94].

6.2.2.2.4 MdtEF

MdtEF is an antibiotic efflux pump complex of the RND type. It provides resistance to fluoroquinolones, macrolides, and penams [82].

6.2.2.3 Other members of the RND (resistance-nodulation-cell division superfamily)

6.2.2.3.1 MexAB-OprM efflux system

MexAB-OprM efflux system is an antibiotic efflux pump complex of the RND type. It provides resistance to multiple antibiotics, including aminocoumarins, carbapenems, cephalosporins, cephamycins, diaminopyrimidines, fluoroquinolones, macrolides, monobactams, penams, phenicols, peptides, sulfonamides, and tetracyclines [95, 96].

6.2.2.3.2 MexCD-OprJ with type A NfxB mutation

MexCD-OprJ with type A NfxB mutation is an antibiotic efflux pump complex of the RND type. It provides resistance to the aminocoumarins, cephalosporins, diaminopyrimidines, fluoroquinolones, macrolides, penams, phenicols, and tetracyclines [97].

6.2.2.3.3 MexCD-OprJ with type B NfxB mutation

MexCD-OprJ with type B NfxB mutation is an antibiotic efflux pump complex of the RND type. It provides resistance to the aminocoumarins, aminoglycosides, cephalosporins, diaminopyrimidines, fluoroquinolones, macrolides, penams, phenicols, and tetracyclines [97].

6.2.2.3.4 MexEF-OprN

MexEF-OprN is an antibiotic efflux pump complex RND. It provides resistance to diaminopyrimidines, fluoroquinolones, and phenicols [98].

6.2.2.3.5 MexXY-OprM

MexXY-OprM is an antibiotic efflux pump complex RND. It provides resistance to the acridine dye, aminoglycosides, carbapenems, cephalosporins, cephamycins, fluoroquinolones, macrolides, penams, phenicols, and tetracyclines [96, 99, 100].

6.2.2.3.6 CmeABC

CmeABC is an antibiotic efflux pump complex RND. It provides resistance to cephalosporins, fluoroquinolones, fusidic acid, and macrolides [101, 102].

6.2.2.3.7 AdeIJK

AdeIJK is an antibiotic efflux pump complex RND. It provides resistance to carbapenems, cephalosporins, diaminopyrimidines, fluoroquinolones, lincosamides, macrolides, penems, phenicols, rifamycins, and tetracyclines [103].

6.2.2.3.8 AdeABC

AdeABC is an antibiotic efflux pump complex RND. It provides resistance to glycylcyclines and tetracyclines [104, 105].

6.2.2.3.9 AdeL

AdeL is an antibiotic efflux pump complex RND. It provides resistance to fluoroquinolones and tetracyclines [106].

6.2.2.3.10 SmeDEF

SmeDEF is an antibiotic efflux pump complex RND. It provides resistance to fluoroquinolones, macrolides, phenicols, and tetracyclines [107].

Other molecular complexes associated with decreasing the intracytoplasmic concentration of antibiotics in Gram-negative bacteria include:

6.2.2.4 Members of the MFS (major facilitator superfamily) in Gram-negative bacteria

6.2.2.4.1 EmrAB-TolC

EmrAB-TolC is an antibiotic efflux pump belonging to MFS. It provides resistance to fluoroquinolones [108].

6.2.2.4.2 MdfA

MdfA is an antibiotic efflux pump belonging to MFS. It provides resistance to benzalkonium chloride, fluoroquinolones, rhodamine, and tetracyclines [109, 110].

6.2.2.5 Other Gram-negative mechanisms

Other molecular complexes associated with decreasing the intracytoplasmic concentration of antibiotics in Gram-negative bacteria include:

6.2.2.5.1 Porin OprF

The OprF porin channel is permeable to quinolones and other antibiotics, promoting its outflow and decreasing intracytoplasmic concentration and consequently is a mechanism of antibiotic resistance for the bacteria [111, 112].

6.3 Plasmid-mediated quinolone resistance genes (plasmids that protect cells from the lethal effects of quinolones)

In 1998 at the University of Alabama, from the isolation of *Klebsiella pneumoniae* from a urine sample, Martinez et al. managed to identify a plasmid they named *pMG252*. They demonstrated that this plasmid induced bacterial resistance to fluoroquinines and nalidixic acid. This resistance phenomenon could be induced in a variety of bacteria deficient in outer-membrane porins. They also described that this plasmid promoted the acceleration of resistance development and its propagation. The gene responsible for this resistance was called *qnr*, later it became *qnrA* [113, 114].

In 2002, Tran and Jacoby, working with the qnr plasmid, managed to identify an integron-like environment upstream from qacEΔ1 and sulI. The product obtained from this gene was a 218-aa protein called QnrA. This protein belonging to the pentapeptide repeat family shared sequence homology with the immunity protein McbG. Previous studies suggested that McbG protects DNA gyrase from the action of various genotoxic chemicals [115].

Based on the mechanism of action of quinolones (the inhibition of topoisomerases I and IV) and the similarity of QnrA to McbG, Tran and Jacoby determined the ability of QnrA to induce resistance against quinolones by topoisomerase protection [115].

In 2005, two independent teams managed to determine the same activity as QnrA for two other proteins identified as QnrB [116] and QnrS [117].

Subsequent studies of the qnrA plasmid found that this plasmid was able to promote greater resistance than expected and that is how, in 2006, Ari Robicsek et al. discovered another mechanism of action of resistance to quinolones mediated by the enzymatic action of aminoglycoside acetyltransferase, AAC(6′)-Ib-cr. They also reported that the quinolone resistance mechanism was determined by reduction of the activity of ciprofloxacin by N-acetylation at the amino nitrogen on its piperazinyl substituent [118].

In 2007, three groups of researchers separately demonstrated another resistance mechanism encoded by plasmids. These works, in correlation with Martinez's works, involve quinolone efflux pumps mediated by plasmids QepA [119, 120] and OqxAB [121].

In summary, there are three mechanisms for PMQR:

1. The plasmid genes *qnrA*, *qnrB*, *qnrC*, *qnrD*, *qnrS,* and *qnrVC* encode proteins from the pentapeptide repeat family that protects DNA gyrase and topoisomerase IV from quinolone inhibition. The *qnr* genes are generally associated with mobilizing or transposable elements in plasmids and are often incorporated into sul1-type integrons.

2. The second mechanism mediated by plasmids involves acetylation of quinolones with an appropriate amino nitrogen target by a variant of the common aminoglycoside acetyltransferase AAC(6′)-Ib-cr.

3. Improved outflow produced by plasmid genes for QepAB and OqxAB pumps.

7. Concluding remarks

Bacterial resistance to antibiotics is a serious problem worldwide and offers the bleakest outlook and prognosis. The number of reports of isolation of multiresistant strains is increasing, including antibiotics of the latest generation or exclusive intra-hospital use. In this sense, isolates of strains resistant to practically all members of the quinolone family have been reported.

The implementation of appropriate practices in the use of antibiotics plays an important role in the fight against this serious global problem. The proper management of antibiotics must include limiting their use in the livestock, agricultural, and food industries; as well as the correct medical prescription, avoiding self-medication, and always seeking adherence to the full antibiotical treatment scheme.

The knowledge of the molecular mechanisms associated with resistance to quinolones and other antibiotics offers us great possibilities for molecular epidemiological monitoring of the emergence of new resistant strains, as well as their distribution. This knowledge offers the pharmaceutical industry the tools for the development of new drugs. It is important to consider that the development time of new drugs is exceeded by the speed of appearance of new resistant strains.

Author details

Sandra Georgina Solano-Gálvez[1], María Fernanda Valencia-Segrove[2], María José Ostos Prado[2], Ana Berenice López Boucieguez[2], Diego Abelardo Álvarez-Hernández[2] and Rosalino Vázquez-López[2]*

1 Departamento de Microbiología, Facultad de Medicina de la Universidad Nacional Autónoma de México, Mexico

2 Departamento de Microbiología, Centro de Investigación en Ciencias de la Salud (CICSA), Facultad de Ciencias de la Salud, Universidad Anáhuac México Norte, Mexico

*Address all correspondence to: rosalino.vazquez@anahuac.mx

References

[1] WHO. WHO | High Levels of Antibiotic Resistance Found Worldwide, New Data Shows [Internet] [cited 01 February 2020]. Available from: https://www.who.int/mediacentre/news/releases/2018/antibiotic-resistance-found/en/

[2] Biggest Threats and Data | Antibiotic/Antimicrobial Resistance | CDC 2020 [Internet] [cited 01 February 2020]. Available from: https://www.cdc.gov/drugresistance/biggest-threats.html

[3] Magiorakos A-P, Srinivasan A, Carey RB, Carmeli Y, Falagas ME, Giske CG, et al. Multidrug-resistant, extensively drug-resistant and pandrug-resistant bacteria: An international expert proposal for interim standard definitions for acquired resistance. Clinical Microbiology and Infection [Internet]. 2012;**18**(3):268-281 [cited 11 June 2019]. Available from: http://www.ncbi.nlm.nih.gov/pubmed/21793988

[4] Vázquez-López R, Rivero Rojas O, Ibarra Moreno A, Urrutia Favila JE, Peña Barreto A, Ortega Ortuño GL, et al. Antibiotic-resistant septicemia in pediatric oncology patients associated with post-therapeutic neutropenic fever. Antibiotics [Internet]. 2019;**8**(3):106. Available from: https://www.mdpi.com/2079-6382/8/3/106

[5] Malekzadegan Y, Rastegar E, Moradi M, Heidari H, Sedigh E-SH. Prevalence of quinolone-resistant uropathogenic *Escherichia coli* in a tertiary care hospital in South Iran. Infection and Drug Resistance. 2019;**12**:1683-1689

[6] Kim ES, Hooper DC. Clinical importance and epidemiology of quinolone resistance. Infection and Chemotherapy. 2014;**46**:226-238

[7] TDM P, Ziora ZM, MAT B. Quinolone antibiotics. MedChemComm. 2019;**10**:1719-1739

[8] Emmerson AM. The quinolones: Decades of development and use. The Journal of Antimicrobial Chemotherapy. 2003;**51**(90001):13-20

[9] Piddock LJV. Fluoroquinolone resistance. British Medical Journal. 1998;**317**:1029-1030

[10] Redgrave LS, Sutton SB, Webber MA, Piddock LJV. Fluoroquinolone resistance: Mechanisms, impact on bacteria, and role in evolutionary success. Trends in Microbiology. 2014;**22**:438-445

[11] EARS-Net. Antimicrobial resistance surveillance in Europe; 2015

[12] Ecdc. SURVEILLANCE REPORT. Surveillance of antimicrobial resistance in Europe 2018 [Internet]. 2018 [cited 07 February 2020]. Available from: www.ecdc.europa.eu

[13] Appelbaum PC, Hunter PA. The fluoroquinolone antibacterials: Past, present and future perspectives. International Journal of Antimicrobial Agents. 2000;**16**:5-15

[14] Nalidixic Acid—PubChem [Internet] [cited 12 February 2020]. Available from: https://pubchem.ncbi.nlm.nih.gov/substance/144075330

[15] Aldred KJ, Kerns RJ, Osheroff N. Mechanism of quinolone action and resistance. Biochemistry. 2014;**53**:1565-1574

[16] Kocsis B, Domokos J, Szabo D. Chemical structure and pharmacokinetics of novel quinolone agents represented by avarofloxacin, delafloxacin, finafloxacin, zabofloxacin and nemonoxacin. Annals of Clinical Microbiology and Antimicrobials. 2016b;**15**(1). DOI: 10.1186/s12941-016-0150-4

[17] PubChem. Ciprofloxacin—PubChem [Internet] [cited 12 February 2020].

Available from: https://pubchem.ncbi.nlm.nih.gov/substance/375668987

[18] Ezelarab H, Abbas SH, Hassan HA, Abuo-Rahma G. Recent updates of fluoroquinolones as antibacterial agents. Archiv der Pharmazie. 2018;**351**(9):e1800141. DOI: 10.1002/ardp.201800141

[19] Grepafloxacin–PubChem [Internet] [cited 12 February 2020]. Available from: https://pubchem.ncbi.nlm.nih.gov/substance/318161242

[20] Moxifloxacin—PubChem [Internet] [cited 12 February 2020]. Available from: https://pubchem.ncbi.nlm.nih.gov/substance/318161711

[21] Quinolones and the Clinical Laboratory | HAI | CDC [Internet]. [cited 30 January 2020]. Available from: https://www.cdc.gov/hai/settings/lab/quinolones-clinical-laboratory.html

[22] Correia S, Poeta P, Hébraud M, Capelo JL, Igrejas G. Mechanisms of quinolone action and resistance: Where do we stand? Journal of Medical Microbiology. 2017;**66**:551-559

[23] Blanca-López N, Andreu I, Torrés Jaén MJ. Hypersensitivity reactions to quinolones. Current Opinion in Allergy and Clinical Immunology. 2011;**11**:285-291

[24] Doña I, Moreno E, Pérez-Sánchez N, Andreu I, Hernández Fernandez de Rojas D, Torres MJ. Update on quinolone allergy. Current Allergy and Asthma Reports. 2017;**17**:1

[25] Ashley RE, Dittmore A, McPherson SA, Turnbough CL, Neuman KC, Osheroff N. Activities of gyrase and topoisomerase IV on positively supercoiled DNA. Nucleic Acids Research. 2017;**45**(16):9611-9624

[26] Nitiss JL. Roles of DNA topoisomerases in chromosomal replication and segregation. Advances in Pharmacology. 1994;**29A**:103-134

[27] Collin F, Karkare S, Maxwell A. Exploiting bacterial DNA gyrase as a drug target: Current state and perspectives. Applied Microbiology and Biotechnology. 2011;**92**:479-497

[28] Drlica K, Zhao X. DNA gyrase, topoisomerase IV, and the 4-quinolones. Microbiology and Molecular Biology Reviews. 1997;**61**(3):377-392

[29] Ince D, Zhang X, Silver LC, Hooper DC. Dual targeting of DNA gyrase and topoisomerase IV: Target interactions of garenoxacin (BMS-284756, T-3811ME), a new desfluoroquinolone. Antimicrobial Agents and Chemotherapy. 2002;**46**(11):3370-3380

[30] Reece RJ, Maxwell A, Wang JC. DNA gyrase: Structure and function. Critical Reviews in Biochemistry and Molecular Biology. 1991;**26**(3-4):335-375

[31] Morais Cabral JH, Jackson AP, Smith CV, Shikotra N, Maxwell A, Liddington RC. Crystal structure of the breakage-Reunion domain of DNA gyrase. Nature. 1997;**388**(6645):903-906

[32] Corbett KD, Schoeffler AJ, Thomsen ND, Berger JM. The structural basis for substrate specificity in DNA topoisomerase IV. Journal of Molecular Biology. 2005;**351**(3):545-561

[33] Hooper DC, Jacoby GA. Mechanisms of drug resistance: Quinolone resistance. Annals of the New York Academy of Sciences. 2015;**1354**(1):12-31

[34] Álvarez-Hernández DA, Garza-Mayén GS, Vázquez-López R. Quinolones. Nowadays perspectives and mechanisms of resistance. Revista chilena de infectología. 2015;**32**(5):499-504. DOI: 10.4067/S0716-10182015000600002

[35] Hooper DC, Jacoby GA. Topoisomerase inhibitors: Fluoroquinolone mechanisms of action and resistance. Cold Spring Harbor Perspectives in Medicine. 2016;**6**(9):a025320. DOI: 10.1101/cshperspect.a025320

[36] Wohlkonig A, Chan PF, Fosberry AP, Homes P, Huang J, Kranz M, et al. Structural basis of quinolone inhibition of type IIA topoisomerases and target-mediated resistance. Nature Structural & Molecular Biology. 2010;**17**(9):1152-1153

[37] Laponogov I, Veselkov DA, Crevel IMT, Pan XS, Fisher LM, Sanderson MR. Structure of an "open" clamp type II topoisomerase-DNA complex provides a mechanism for DNA capture and transport. Nucleic Acids Research. 2013;**41**(21):9911-9923

[38] Laponogov I, Sohi MK, Veselkov DA, Pan XS, Sawhney R, Thompson AW, et al. Structural insight into the quinolone-DNA cleavage complex of type IIA topoisomerases. Nature Structural & Molecular Biology. 2009;**16**(6):667-669

[39] Yoshida H, Bogaki M, Nakamura M, Nakamura S. Quinolone resistance-determining region in the DNA gyrase gyrA gene of *Escherichia coli*. Antimicrobial Agents and Chemotherapy. 1990;**34**(6):1271-1272

[40] Breines DM, Ouabdesselam S, Ng EY, Tankovic J, Shah S, Soussy CJ, et al. Quinolone resistance locus nfxD of *Escherichia coli* is a mutant allele of the parE gene encoding a subunit of topoisomerase IV. Antimicrobial Agents and Chemotherapy. 1997;**41**(1):175-179

[41] Yoshida H, Bogaki M, Nakamura M, Yamanaka LM, Nakamura S. Quinolone resistance-determining region in the DNA gyrase gyrB gene of *Escherichia coli*. Antimicrobial Agents and Chemotherapy. 1991;**35**(8):1647-1650

[42] Blanche F, Cameron B, Bernard FX, Maton L, Manse B, Ferrero L, et al. Differential behaviors of Staphylococcus aureus and *Escherichia coli* type II DNA topoisomerases. Antimicrobial Agents and Chemotherapy [Internet]. 1996;**40**(12):2714-2720 [cited 20 January 2020]. Available from: http://www.ncbi.nlm.nih.gov/pubmed/9124828

[43] Pan XS, Fisher LM. Targeting of DNA gyrase in Streptococcus pneumoniae by sparfloxacin: selective targeting of gyrase or topoisomerase IV by quinolones. Antimicrobial Agents and Chemotherapy [Internet]. 1997;**41**(2):471-474. Available from: http://www.ncbi.nlm.nih.gov/pubmed/9021211

[44] Rahman T, Yarnall B, Doyle DA. Efflux drug transporters at the forefront of antimicrobial resistance. European Biophysics Journal. 2017;**46**:647-653

[45] Ghai I, Ghai S. Understanding antibiotic resistance via outer membrane permeability. Infection and Drug Resistance. 2018;**11**:523-530

[46] Grkovic S, Brown MH, Skurray RA. Regulation of bacterial drug export systems. Microbiology and Molecular Biology Reviews. 2002;**66**(4):671-701

[47] Lekshmi M, Ammini P, Adjei JM, Sanford L, Shrestha U, Kumar S, et al. Modulation of antimicrobial efflux pumps of the major facilitator superfamily in *Staphylococcus aureus*. AIMS Microbiology. 2018;**4**(1):1-18

[48] Sharma A, Gupta VK, Pathania R. Efflux pump inhibitors for bacterial pathogens: From bench to bedside. Indian Journal of Medical Research. 2019;**149**:129-145

[49] JMA B, Richmond GE, Piddock LJV. Multidrug efflux pumps in Gram-negative bacteria and their role in antibiotic resistance. Future Microbiology [Internet]. 2014;**9**(10):1165-1177 [cited 24 January 2020]. Available from: http://www.ncbi.nlm.nih.gov/pubmed/25405886

[50] Costa SS, Viveiros M, Amaral L, Couto I. Multidrug efflux pumps in *Staphylococcus aureus*: An update. The Open Microbiology Journal. 2013;**7**(1):59-71

[51] Elbaz Y, Salomon T, Schuldiner S. Identification of a glycine motif required for packing in EmrE, a multidrug transporter from *Escherichia coli*. The Journal of Biological Chemistry. 2008;**283**(18):12276-12283

[52] Bay DC, Turner RJ. Membrane composition influences the topology bias of bacterial integral membrane proteins. Biochimica et Biophysica Acta—Biomembranes. 2013;**1828**(2):260-270

[53] Bay DC, Rommens KL, Turner RJ. Small multidrug resistance proteins: A multidrug transporter family that continues to grow [Internet]. Biochimica et Biophysica Acta—Biomembranes. 2008;**1778**:1814-1838 [cited 15 April 2020]. Available from: http://www.ncbi.nlm.nih.gov/pubmed/17942072

[54] Truong-Bolduc QC, Strahilevitz J, Hooper DC. NorC, a new efflux pump regulated by MgrA of Staphylococcus aureus. Antimicrobial Agents and Chemotherapy. 2006;**50**(3):1104-1107

[55] Ubukata K, Itoh-Yamashita N, Konno M. Cloning and expression of the norA gene for fluoroquinolone resistance in *Staphylococcus aureus*. Antimicrobial Agents and Chemotherapy. 1989;**33**(9):1535-1539

[56] Yoshida H, Bogaki M, Nakamura S, Ubukata K, Konno M. Nucleotide sequence and characterization of the Staphylococcus aureus norA gene, which confers resistance to quinolones. Journal of Bacteriology. 1990;**172**(12):6942-6949

[57] Neyfakh AA, Borsch CM, Kaatz GW. Fluoroquinolone resistance protein NorA of *Staphylococcus aureus* is a multidrug efflux transporter. Antimicrobial Agents and Chemotherapy [Internet]. 1993;**37**(1):128-129 [cited 24 January 2020]. Available from: http://www.ncbi.nlm.nih.gov/pubmed/8431010

[58] Costa SS, Sobkowiak B, Parreira R, Edgeworth JD, Viveiros M, Clark TG, et al. Genetic diversity of norA, coding for a main efflux pump of *Staphylococcus aureus*. Frontiers in Genetics. 2019;**9**:710. DOI: 10.3389/fgene.2018.00710

[59] Ding Y, Onodera Y, Lee JC, Hooper DC. NorB, an efflux pump in *Staphylococcus aureus* strain MW2, contributes to bacterial fitness in abscesses. Journal of Bacteriology. 2008;**190**(21):7123-7129

[60] Kaatz GW, Thyagarajan RV, Seo SM. Effect of promoter region mutations and mgrA overexpression on transcription of norA, which encodes a *Staphylococcus aureus* multidrug efflux transporter. Antimicrobial Agents and Chemotherapy. 2005;**49**(1):161-169

[61] Truong-Bolduc QC, Hooper DC. The transcriptional regulators NorG and MgrA modulate resistance to both quinolones and β-lactams in *Staphylococcus aureus*. Journal of Bacteriology. 2007;**189**(8):2996-3005

[62] Truong-Bolduc QC, Zhang X, Hooper DC. Characterization of NorR protein, a multifunctional regulator of norA expression in *Staphylococcus*

aureus. Journal of Bacteriology [Internet]. 2003;(10):185, 3127-3138 [cited 25 January 2020] Available from: http://www.ncbi.nlm.nih.gov/pubmed/12730173

[63] Truong-Bolduc QC, Bolduc GR, Okumura R, Celino B, Bevis J, Liao CH, et al. Implication of the NorB efflux pump in the adaptation of *Staphylococcus aureus* to growth at acid ph and in resistance to moxifloxacin. Antimicrobial Agents and Chemotherapy. 2011;**55**(7):3214-3219

[64] Truong-Bolduc QC, Hsing LC, Villet R, Bolduc GR, Estabrooks Z, Florent Taguezem G, et al. Reduced aeration affects the expression of the NorB efflux pump of *Staphylococcus aureus* by posttranslational modification of MgrA. Journal of Bacteriology. 2012;**194**(7):1823-1834

[65] Chen PR, Bae T, Williams WA, Duguid EM, Rice PA, Schneewind O, et al. An oxidation-sensing mechanism is used by the global regulator MgrA in *Staphylococcus aureus*. Nature Chemical Biology. 2006;**2**(11):591-595

[66] Truong-Bolduc QC, Dunman PM, Eidem T, Hooper DC. Transcriptional profiling analysis of the global regulator NorG, a GntR-like protein of *Staphylococcus aureus*. Journal of Bacteriology. 2011;**193**(22):6207-6214

[67] Huang J, O'Toole PW, Shen W, Amrine-Madsen H, Jiang X, Lobo N, et al. Novel chromosomally encoded multidrug efflux transporter MdeA in *Staphylococcus aureus*. Antimicrobial Agents and Chemotherapy. 2004;**48**(3):909-917

[68] Yamada Y, Shiota S, Mizushima T, Kuroda T, Tsuchiya T. Functional gene cloning and characterization of MdeA, a multidrug efflux pump from *Staphylococcus aureus*. Biological & Pharmaceutical Bulletin. 2006;**29**(4):801-804

[69] Yamada Y, Hideka K, Shiota S, Kuroda T, Tsuchiya T. Gene cloning and characterization of SdrM, a chromosomally-encoded multidrug efflux pump, from *Staphylococcus aureus*. Biological and Pharmaceutical Bulletin [Internet]. 2006;**29**(3):554-556 [cited 26 January 2020] Available from: http://joi.jlc.jst.go.jp/JST.JSTAGE/bpb/29.554?from=CrossRef

[70] Nakaminami H, Noguchi N, Sasatsu M. Fluoroquinolone efflux by the plasmid-mediated multidrug efflux pump QacB variant QacBIII in *Staphylococcus aureus*. Antimicrobial Agents and Chemotherapy. 2010;**54**(10):4107-4111

[71] Dabul ANG, Avaca-Crusca JS, Van Tyne D, Gilmore MS, Camargo ILBC. Resistance in in vitro selected tigecycline-resistant methicillin-resistant *Staphylococcus aureus* sequence type 5 is driven by mutations in mepR and mepA genes. Microbial Drug Resistance. 2018;**24**(5):519-526

[72] McAleese F, Petersen P, Ruzin A, Dunman PM, Murphy E, Projan SJ, et al. A novel MATE family efflux pump contributes to the reduced susceptibility of laboratory-derived *Staphylococcus aureus* mutants to tigecycline. Antimicrobial Agents and Chemotherapy. 2005;**49**(5):1865-1871

[73] Kaatz GW, McAleese F, Seo SM. Multidrug resistance in *Staphylococcus aureus* due to overexpression of a novel multidrug and toxin extrusion (MATE) transport protein. Antimicrobial Agents and Chemotherapy. 2005;**49**(5):1857-1864

[74] Nikaido H. Structure and mechanism of RND-type multidrug efflux pumps. Advances in Enzymology and Related Areas of Molecular Biology. 2010;**77**(1):1-60

[75] Anes J, McCusker MP, Fanning S, Martins M. The ins and outs of RND

efflux pumps in *Escherichia coli*. Frontiers in Microbiology. 2015;**6**:587. DOI: 10.3389/fmicb.2015.00587

[76] Venter H, Mowla R, Ohene-Agyei T, Ma S. RND-type drug efflux pumps from Gram-negative bacteria: Molecular mechanism and inhibition. Frontiers in Microbiology. 2015;**6**:377. DOI: 10.3389/fmicb.2015.00377

[77] Kim EH, Nies DH, McEvoy MM, Rensing C. Switch or funnel: How RND-type transport systems control periplasmic metal homeostasis. Journal of Bacteriology. 2011;**193**:2381-2387

[78] Tikhonova EB, Zgurskaya HI. AcrA, AcrB, and TolC of *Escherichia coli* form a stable intermembrane multidrug efflux complex. The Journal of Biological Chemistry. 2004;**279**(31):32116-32124

[79] Pos KM. Drug transport mechanism of the AcrB efflux pump. Biochimica et Biophysica Acta, Proteins and Proteomics. 2009;**1794**:782-793

[80] Rosenberg EY, Ma D, Nikaido H. AcrD of *Escherichia coli* is an aminoglycoside efflux pump. Journal of Bacteriology. 2000;**182**(6):1754-1756

[81] Elkins CA, Nikaido H. Substrate specificity of the RND-type multidrug efflux pumps AcrB and AcrD of *Escherichia coli* is determined predominantly by two large periplasmic loops. Journal of Bacteriology [Internet]. 2002;**184**(23):6490-6498 [cited 31 January 2020]. Available from: http://www.ncbi.nlm.nih.gov/pubmed/12426336

[82] Nishino K, Yamaguchi A. Role of histone-like protein H-NS in multidrug resistance of *Escherichia coli*. Journal of Bacteriology. 2004;**186**(5):1423-1429

[83] Baranova N, Nikaido H. The baeSR two-component regulatory system activates transcription of the yegMNOB (mdtABCD) transporter gene cluster in *Escherichia coli* and increases its resistance to novobiocin and deoxycholate. Journal of Bacteriology [Internet]. 2002;**184**(15):4168-4176 [cited 31 January 2020]. Available from: http://www.ncbi.nlm.nih.gov/pubmed/12107134

[84] Nagakubo S, Nishino K, Hirata T, Yamaguchi A. The putative response regulator BaeR stimulates multidrug resistance of *Escherichia coli* via a novel multidrug exporter system, MdtABC. Journal of Bacteriology [Internet]. 2002;**184**(15):4161-4167 [cited 27 January 2020]. Available from: http://www.ncbi.nlm.nih.gov/pubmed/12107133

[85] Hirakawa H, Inazumi Y, Senda Y, Kobayashi A, Hirata T, Nishino K, et al. N-acetyl-D-glucosamine induces the expression of multidrug exporter genes, mdtEF, via catabolite activation in *Escherichia coli*. Journal of Bacteriology. 2006;**188**(16):5851-5858

[86] Zhang Y, Xiao M, Horiyama T, Zhang Y, Li X, Nishino K, et al. The multidrug efflux pump MdtEF protects against nitrosative damage during the anaerobic respiration in *Escherichia coli*. The Journal of Biological Chemistry. 2011;**286**(30):26576-26584

[87] Du D, Wang Z, James NR, Voss JE, Klimont E, Ohene-Agyei T, et al. Structure of the AcrAB-TolC multidrug efflux pump. Nature. 2014;**509**(7501):512-515

[88] Yu EW, McDermott G, Zgurskaya HI, Nikaido H, Koshland DE. Structural basis of multiple drug-binding capacity of the AcrB multidrug efflux pump. Science. 2003;**300**(5621):976-980

[89] Murakami S, Nakashima R, Yamashita E, Yamaguchi A. Crystal structure of bacterial multidrug efflux transporter AcrB. Nature. 2002;**419**(6907):587-593

[90] Poole K. Efflux-mediated multiresistance in Gram-negative

bacteria. Clinical Microbiology and Infection. 2004;**10**:12-26

[91] Lau SY, Zgurskaya HI. Cell division defects in *Escherichia coli* deficient in the multidrug efflux transporter AcrEF-TolC. Journal of Bacteriology. 2005;**187**(22):7815-7825

[92] Poole K. Efflux-mediated resistance to fluoroquinolones in gram-negative bacteria. Antimicrobial Agents and Chemotherapy. 2000;**44**:2233-2241

[93] Nishino K, Nikaido E, Yamaguchi A. Regulation of multidrug efflux systems involved in multidrug and metal resistance of *Salmonella enterica* serovar typhimurium. Journal of Bacteriology. 2007;**189**(24):9066-9075

[94] Nagakubo S, Nishino K, Hirata T, Yamaguchi A. The putative response regulator BaeR stimulates multidrug resistance of *Escherichia coli* via a novel multidrug exporter system, MdtABC. Journal of Bacteriology [Internet]. 2002;**184**(15):4161-4167 [cited 31 January 2020] Available from: http://www.ncbi.nlm.nih.gov/pubmed/12107133

[95] Choudhury D, Ghosh A, Chanda DD, Das TA, Choudhury MD, Paul D, et al. Premature termination of MexR leads to overexpression of MexAB-OprM efflux pump in *Pseudomonas aeruginosa* in a tertiary referral hospital in India. PLoS One. 2016;**11**(2):1

[96] Li XZ, Zhang L, Poole K. Interplay between the MexA-MexB-OprM multidrug efflux system and the outer membrane barrier in the multiple antibiotic resistance of *Pseudomonas aeruginosa*. The Journal of Antimicrobial Chemotherapy. 2000;**45**(4):433-436

[97] Masuda N, Gotoh N, Ohya S, Nishino T. Quantitative correlation between susceptibility and OprJ production in NfxB mutants of *Pseudomonas aeruginosa*. Antimicrobial Agents and Chemotherapy. 1996;**40**(4):909-913

[98] Richardot C, Juarez P, Jeannot K, Patry I, Plésiat P, Llanes C. Amino acid substitutions account for most mexS alterations in clinical nfxC mutants of *Pseudomonas aeruginosa*. Antimicrobial Agents and Chemotherapy. 2016;**60**(4):2302-2310

[99] Mine T, Morita Y, Kataoka A, Mizushima T, Tsuchiya T. Expression in *Escherichia coli* of a new multidrug efflux pump, MexXY, from *Pseudomonas aeruginosa*. Antimicrobial Agents and Chemotherapy. 1999;**43**(2):415-417

[100] Hocquet D, Vogne C, El Garch F, Vejux A, Gotoh N, Lee A, et al. MexXy-OprM efflux pump is necessary for adaptive resistance of *Pseudomonas aeruginosa* to aminoglycosides. Antimicrobial Agents and Chemotherapy. 2003;**47**(4):1371-1375

[101] Lin J, Akiba M, Sahin O, Zhang Q. CmeR functions as a transcriptional repressor for the multidrug efflux pump CmeABC in *Campylobacter jejuni*. Antimicrobial Agents and Chemotherapy. 2005;**49**(3):1067-1075

[102] Yao H, Shen Z, Wang Y, Deng F, Liu D, Naren G, et al. Emergence of a potent multidrug efflux pump variant that enhances *Campylobacter* resistance to multiple antibiotics. MBio. 2016;**7**(5):1

[103] Damier-Piolle L, Magnet S, Brémont S, Lambert T, Courvalin P. AdeIJK, a resistance-nodulation-cell division pump effluxing multiple antibiotics in *Acinetobacter baumannii*. Antimicrobial Agents and Chemotherapy. 2008;**52**(2):557-562

[104] Ruzin A, Keeney D, Bradford PA. AdeABC multidrug efflux pump is associated with decreased susceptibility to tigecycline in *Acinetobacter*

calcoaceticus-Acinetobacter baumannii complex. The Journal of Antimicrobial Chemotherapy. 2007;**59**(5):1001-1004

[105] Coyne S, Courvalin P, Périchon B. Efflux-mediated antibiotic resistance in Acinetobacter spp. Antimicrobial Agents and Chemotherapy. 2011;**55**:947-953

[106] Coyne S, Rosenfeld N, Lambert T, Courvalin P, Perichon B. Overexpression of resistance-nodulation-cell division pump AdeFGH confers multidrug resistance in *Acinetobacter baumannii*. Antimicrobial Agents and Chemotherapy [Internet]. 2010;**54**(10):4389-4393 [cited 08 March 2019] Available from: http://www.ncbi.nlm.nih.gov/pubmed/20696879

[107] Zhang L, Li XZ, Poole K. SmeDEF multidrug efflux pump contributes to intrinsic multidrug resistance in *Stenotrophomonas maltophilia*. Antimicrobial Agents and Chemotherapy. 2001;**45**(12):3497-3503

[108] Lomovskaya O, Lewis K, Matin A. EmrR is a negative regulator of the *Escherichia coli* multidrug resistance pump emrAB. Journal of Bacteriology. 1995;**177**(9):2328-2334

[109] Bohn C, Bouloc P. The *Escherichia coli* cmlA gene encodes the multidrug efflux pump Cmr/MdfA and is responsible for isopropyl-β-D-thiogalactopyranoside exclusion and spectinomycin sensitivity. Journal of Bacteriology. 1998;**180**(22):6072-6075

[110] Zhao Y, Heng J, Zhao Y, Liu M, Liu Y, Fan J, et al. Substrate-bound structure of the *E. coli* multidrug resistance transporter MdfA. Cell Research. 2015;**25**(9):1060-1073

[111] Nestorovich EM, Sugawara E, Nikaido H, Bezrukov SM. Pseudomonas aeruginosa porin OprF. Properties of the channel. The Journal of Biological Chemistry. 2006;**281**(24):16230-16237

[112] Nikaido H, Nikaido K, Harayama S. Identification and characterization of porins in *Pseudomonas aeruginosa*. The Journal of Biological Chemistry. 1991;**266**(2):770-779

[113] Martínez-Martínez L, Pascual A, Jacoby GA. Quinolone resistance from a transferable plasmid. Lancet. 1998;**351**(9105):797-799

[114] Jacoby GA. Mechanisms of resistance to quinolones. Clinical Infectious Diseases [Internet]. 2005;**41**(Supplement_2):S120-S126. [cited 17 January 2020]. Available from: http://academic.oup.com/cid/article/41/Supplement_2/S120/307501/Mechanisms-of-Resistance-to-Quinolones

[115] Tran JH, Jacoby GA. Mechanism of plasmid-mediated quinolone resistance. Proceedings of the National Academy of Sciences of the United States of America. 2002;**99**(8):5638-5642

[116] Jacoby GA, Walsh KE, Mills DM, Walker VJ, Oh H, Robicsek A, et al. qnrB, another plasmid-mediated gene for quinolone resistance. Antimicrobial Agents and Chemotherapy. 2006;**50**(4):1178-1182

[117] Hata M, Suzuki M, Matsumoto M, Takahashi M, Sato K, Ibe S, et al. Cloning of a novel gene for quinolone resistance from a transferable plasmid in *Shigella flexneri* 2b. Antimicrobial Agents and Chemotherapy. 2005;**49**(2):801-803

[118] Robicsek A, Strahilevitz J, Jacoby GA, Macielag M, Abbanat D, Chi HP, et al. Fluoroquinolone-modifying enzyme: A new adaptation of a common aminoglycoside acetyltransferase. Nature Medicine. 2006;**12**(1):83-88

[119] Yamane K, Wachino JI, Suzuki S, Kimura K, Shibata N, Kato H, et al. New plasmid-mediated fluoroquinolone efflux pump, QepA, found in an

Escherichia coli clinical isolate. Antimicrobial Agents and Chemotherapy. 2007;**51**(9):3354-3360

[120] Périchon B, Courvalin P, Galimand M. Transferable resistance to aminoglycosides by methylation of G1405 in 16S rRNA and to hydrophilic fluoroquinolones by QepA-mediated efflux in *Escherichia coli*. Antimicrobial Agents and Chemotherapy. 2007;**51**(7):2464-2469

[121] Hansen LH, Jensen LB, Sørensen HI, Sørensen SJ. Substrate specificity of the OqxAB multidrug resistance pump in *Escherichia coli* and selected enteric bacteria. Journal of Antimicrobial Chemotherapy [Internet]. 2007;**60**(1):145-147. [cited 30 January 2020]. Available from: http://academic.oup.com/jac/article/60/1/145/731003/Substrate-specificity-of-the-OqxAB-multidrug

Chapter 3

Antimicrobial Resistance in *Pseudomonas aeruginosa*: A Concise Review

Swaraj Mohanty, Bighneswar Baliyarsingh and Suraja Kumar Nayak

Abstract

Pseudomonas aeruginosa is one of the common species responsible for an array of diseases in the respiratory tract, gastrointestinal tract, urinary tract, bones, joints and different systemic infections of normal and immunocompromised patients as well. It exhibits resistance to a wide variety of antimicrobial agents and expresses diverse molecular epidemiology to various established classes of antibiotics including β-lactams, fluoroquinolones, tetracycline and aminoglycosides. Despite the low permeability, hydrophilicity and nonspecific behavior of the outer membrane to small molecular transport, it is inadequate to explain the degree of resistance in *P. aeruginosa*. The resistance mechanism of *P. aeruginosa* against various chemical agents is due to the complex chromosomally encoded genes. Different strains of *P. aeruginosa* having the inherent capacity for biofilm formation, further boosts the resistance under various environmental factors. This chapter explains pathogenicity, mode and types of resistance of *P. aeruginosa*, its impact on the economy and available remediation/reduction measures and treatments.

Keywords: *Pseudomonas aeruginosa*, quorum sensing, adaptive resistance, acquired resistance, intrinsic resistance, efflux system

1. Introduction

Pseudomonas aeruginosa, a Gram-negative pathogen usually found in the hospital, plays a crucial role for nosocomial infection and are also responsible for acute and chronic infection. *P. aeruginosa* is ubiquitous in nature and shows a great susceptibility against various classes of antibiotics [1]. The bacteria get colonize on any surface that contains water and multiply rapidly, carry out all the metabolic functions for growth and development which is an association of complex matrix known as a biofilm [2, 3]. The study predicted that a biofilm makes the bacteria more susceptible in the conditions like antibiotics, exposure in UV light and salinity [4]. Further understanding of the pathogenesis and resistance mechanism is a diverse area of investigation. Due to the complex biofilm forming ability, *Pseudomonas* species shows a great resistivity to various classes of antibiotics which are used to persistently overcome the microbial infection. The occurrence of *Pseudomonas* species in the hospitals helps to form the biofilms on the medical instruments (surface only) and other similar devices along with the implants in the patients [5, 6].

Pseudomonas species are used as a model organism for the study of biochemical mechanisms responsible for the susceptibility of the pathovars against a wide variety of antibiotics groups like amikacin, gentamicin, carbapenem, ofloxacin, ciprofloxacin, tigecycline, tobramycin and norfloxacin [7, 8].

The development of resistance by the pathogenic *Pseudomonas* species devise a major problem in the bacterial diversity by altering the genome sequences and the expression of proteins that ultimately improves the resistance of the pathovars [9, 10]. Various biochemical pathways and channel protein functions are affected due to the resistance of the bacteria [11, 12]. At this alarming stage of the scenario in details studies and prevention measures at an earliest is essential to control the same else in near future it may reach beyond our control. Therefore, the present chapter emphasizes on the infections due to *Pseudomonas aeruginosa*, their mechanism of infection and resistance to various classes of antibiotics.

2. Overview of *Pseudomonas aeruginosa* pathogenesis

The infectious diseases caused by *P. aeruginosa* are sometimes fatal for humans as it is a potential threat to people having less immunity like newborns, diseased persons and veterans. Notably, patients suffering from the diseases like cystic fibrosis, urinary tract infection, burn of the skin, leukemia, HIV-AIDS, diabetes, patients having longer stay in hospital environments and persons having organ transplantation are highly susceptible to *P. aeruginosa*. **Table 1** listed the disease, symptoms and its causes.

Disease caused in humans	Symptoms	Adverse effects on human	References
Bacteremia	Fever, fatigue, chills, joint and muscle pain	Increasing bacterial population in the bloodstream	[13]
Pneumonia, sinusitis	Fever, chills, difficulty in breathing, cough with or without sputum production	Deposition of liquids in the parts of the lungs. Swelling and inflammation of the nasal tract	[14]
Folliculitis	Abscess production in the skin, redness of the skin, draining wounds	Inflammation of the hair follicles by bacteria	[15, 16]
External ear canal Infection (otitis externa)	Ear pain, swelling, itching inside the ear, discharge from the ear, sometimes difficulty in hearing	Frequent showering leads to deposition of water and hence the growth of bacteria takes place at that location	[17, 18]
Corneal inflammation (keratitis)	Redness, pain, swelling, inflammation, pus formation, impaired vision	The bacteria adhere to the lens and other parts of an eye within 24 h of its exposure by its cilia and flagella and forms the biofilm	[19, 20]
Urinary tract infection	Burning with urination, cloudy or bloody urine, strong odor, rectal pain (in male), pelvic pain (in female)	The transfer of bacteria into the urethra	[21]
Diabetic foot	Swelling of foot and ankle, dry cracks in the skin(around the heel), corns or calluses	Tissue damage in the foot and severe pain due to ingrown toenails	[22]

Table 1.
Diseases and symptoms of Pseudomonas aeruginosa *infection.*

SI no.	Strains of *Pseudomonas aeruginosa*	Showing resistance to antibiotic class	Mode of action	References
1.	PA40, PA43	Amikacin	Multi-drug-resistance (MDR)	[24, 25]
2.	ATCC 27853, P2284	Ticarcillin/clavulanate	Production of β-lactamase	[26]
3.	K385	Chloramphenicol and norfloxacin	Overexpression of *mexC-MexD-OprJ* operon	[27]
4.	PA-M4	Ciprofloxacin	Overexpression of *MexEF-OprN* operon	[28]
5.	OCR1	Gentamicin	Overexpression of *MexAB-OprM* operon	[28]
6.	PAO4222	Carbapenem (imipenem and meropenem)	Loss of porin channels in the outer membrane, expression of OprD and secreting carbapenem-hydrolyzing metalloenzyme	[29]
7.	PAO4098E	Carbenicillin and tobramycin	Inactivation of aminoglycosides enzyme, ribosomal methyl group transferase enzyme	[27]
8.	PAO1	Tigecycline	Inhibition of *MexXY-OprM* activity	[30]
9.	KG3002	Ofloxacin	Inactivation of *MexC* operon	[31]
10.	KG3000	Ciprofloxacin	Expression of *MexC-MexD-OprJ* operon	[32]
11.	PAO1	Fluroquinolones	DNA gyrase topoisomerase IV activity	[33, 34]
12.	PA1109	Polymyxin E (colistin)	Modification in the LPS layer	[35, 36]
13.	PA124	Tetracyclines	Activation of *MexXY-OprM* efflux pump	[37]
14.	PAO1	Quinolones	Expression of *MexEF-OprN* efflux pump due to mutation of NfxB, NfxC and NalB	[38, 39]
15.	ATTC 27853, K1178	Cephalosporin	Overexpression of *MexAB-OprM* efflux pump due to the NalB mutation	[40]

Table 2.
Antibiotics resistance in different strains of Pseudomonas aeruginosa.

The resistance of *P. aeruginosa* to different aminoglycoside agents show a tremendous threat to public health as well as constrains the therapeutic choice available. The use of multiple drugs against the diseases in a low dose make the *P. aeruginosa* strains more resistant to a wide range of antibiotics [23]. The different strains of *P. aeruginosa* showing resistance to various antibiotic classes along with the pathway of resistant have been demonstrated in **Table 2**.

3. Pathogenicity of *Pseudomonas aeruginosa*

The virulence property of *P. aeruginosa* is mainly due to the presence of factors like alkaline protease, elastase, pyoverdin, pyocyanin, exotoxins and cytotoxins.

This virulence factors are commonly restricted to immunocompromised patients. The pathovars also produces a kind of exopolysaccharide known as alginate in patients having chronic respiratory infections. These alginate serves as the adhesive on the solid surfaces and also protects the bacteria from unfavorable environmental conditions [41]. The bacteria also produce alginate lyase enzyme which can cleave the polysaccharide into short oligosaccharide units it has been observed that both the biosynthesis and degradation process plays a vital role in the infection process [42, 43]. Presence of extracellular virulence factors and cell surface associated structures promotes its pathogenicity [44, 45].

P. aeruginosa binds to the ganglioside present in the host epithelial surface with the help of lipopolysaccharide and bacterial adhesins (i.e. type-IV pili and flagella). Type-IV also facilitates the bacterial movement along the host cell surface known as "twitching motility" which enhances the development of biofilm [46]. After the attachment to the host cell type III secretion system (T3SS) get activated and makes pore or a channel (i.e. translocon) on the cell membrane by injecting cytotoxic effector proteins into the cytosol of host cell [47, 48]. Mainly four different types of toxins are found in the *P. aeruginosa* sp. i.e. Exoenzymes S, T, U and Y. EXoS, ExoT and ExoU are responsible for N-terminal GTPase-activating proteinase (GAP) activity, C-terminal ADP-ribosyltransferase activity (ADPRT) and adenylate cyclase activity respectively [49]. It has been found that the ExoU is also a potent cytotoxin to cleave the host membrane phospholipid layers i.e. Phospholipase A2 (PLA2) activity. The ExoU initiates the inflammation by secreting the arachidonic acid for activating lipoxygenase and cyclooxygenase pathways and results the production of prostaglandins. *P. aeruginosa* secretes an Exotoxin A which is a type of ADPRT that causes cell death by inhibiting protein synthesis due to suppression of host elongation factor 2(EF2) [50]. The lipase and phospholipase of the bacteria dissolve the surfactant lipids and phospholipids of the host cell membranes. The blue-green pigment pyocyanin develops the oxidative stress in host cells by disrupting the host catalase and electron transport system (ETS) hence suppresses the phagocytosis activity of the host immune system [51].

The type-VI secretion system (T6SS) seen in case of *P. aeruginosa* facilitates the interaction of this pathogen with other organism and provides defence from other bacteria. The H1-, H2- and H3-T6SS are the three distinct T6SS observed in this pathovars. The H1-T6SS is being used for the physiological study of antimicrobial activity [52, 53]. The H2- and H3-T6SS plays dual role in the interaction with both prokaryotic and eukaryotic cell. The production of proteases degrades the covered mucin and complement systems which results the disruption of the tight junctions between the host epithelial cells. Then the bacteria spreads from one cell to others by secreting the phospholipase by damaging the cell membrane [54]. The release of pyocyanin and pyoverdin interfere with the electron transport pathways and redox cycling system of the host cells. LasA and LasB are the two types of elastases produced by *P. aeruginosa,* commonly responsible for the burn wound infection and acute lung infections. The LasA hydrolyze the penta-glycine bridge necessary for the stabilization of the peptidoglycan in the cell wall and the LasB is responsible for the opsonisation of the lung surfactant proteins A and D [55].

4. Resistance for antimicrobials in *Pseudomonas aeruginosa*

A wide group of *P. aeruginosa* strains are resistance to various classes of antibiotics or antibacterial agents that makes it difficult to control the infection. The resistance in *Pseudomonas* species is broadly due to the below detail explained methods studied previously. **Figure 1** explicitly elaborate on various mechanism of

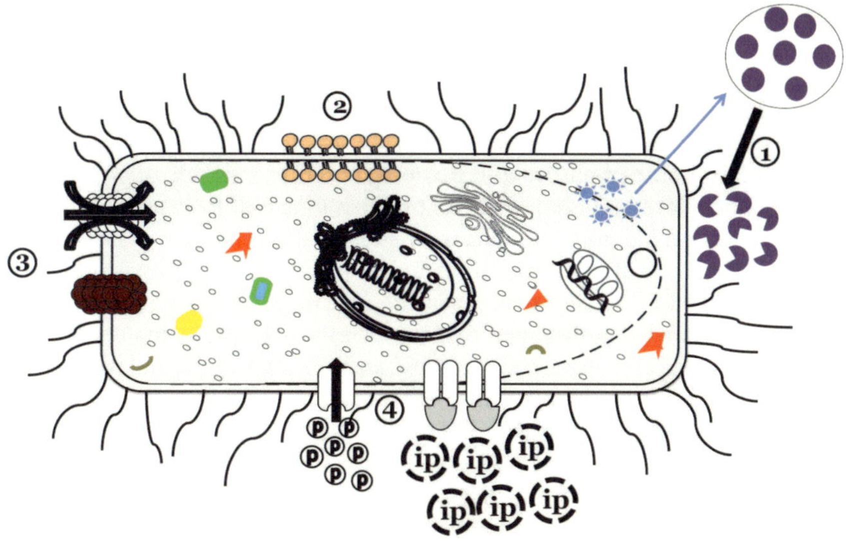

Figure 1.
Resistance of P. aeruginosa *to various antimicrobials as (1) shows the enzymatic modification, (2) impermeability resistance, (3) efflux system and (4) modification in the outer membrane.*

P. aeruginosa resistance. The resistance pattern and mechanism behind the development of resistance in the *Pseudomonas* species are the topic of interest for the researchers as it will help to develop the polyprophylactic procedures and mitigation of infection due to *P. aeruginosa*.

4.1 Enzymatic modification

P. aeruginosa consists of elements generally termed as transposons which induce resistance due to the modification of aminoglycoside enzymes. The infection due to the pathogen is usually combated by various class/groups of aminoglycoside antibiotics like kanamycin, gentamicin, streptomycin, amikacin and neomycin. Previous studies elucidate that, there are three types of enzymatic conformational change which are accountable for the resistance against the bactericidal compounds. These are phosphorylation of aminoglycoside phosphoryl transferase (APH) [56, 57] adenylation of aminoglycoside nucleotidyl transferase (ANT) and acetylation of aminoglycoside acetyl transferase (AAC) [58, 59].

The conformational modification and phosphorylation in the 3′-OH group is carried out by the APH enzyme. APH (3′) family of enzymes shows resistance against streptomycin, butirocin, amikacin, kanamycin and neomycin by encoding the genes such as *aphA* and *hpaA* which are involved in the metabolism of 4-hydroxy-phenylacetic acid (4-HPA). However, APH (2″) shows resistance to tobramycin and gentamycin classes of antibiotics. Due to adenylation of ANT enzymes *P. aeruginosa* increases resistance towards tobramycin, gentamicin, streptomycin, isepamicin and amikacin [60, 61]. The family of enzymes such as ANT (2″), (3″) and (4′) also shows a similar type of resistance in different strains of *P. aeruginosa* isolated from hospitals and intensive care unit (ICU) premises [62]. The *N*-terminal positions (1, 2′, 3 and 6′) of the (AAC) shows the enzymatic acetylation. Amongst various families, AAC (3-I), (3-II) and (3-III) are also resistant to gentamicin, tobramycin and kanamycin antibiotics respectively [63]. Apart from that AAC (6′) family of enzymes contributes to the resistance along with akamicin [64].

4.2 Impermeability resistance

Impermeability to various exocompounds in Gram-negative bacteria is due to lipopolysaccharide (LPS) present in the cell wall. LPS is made up of lipid A, oligosaccharide core and O antigen regions which are linked covalently [65]. The lipid A region is hydrophobic in nature and made up of a disaccharide of glucosamine which is phosphorylated and helps in the anchoring of LPS to the cell membrane. The core oligosaccharide is accumulation of sugar, ethanolamine, phosphate and amino acids and can be divided into inner and outer core. The O antigen is the outer domain of bacterial LPS made up of repeating glycan polymers and attached with the core region. It has been observed that the deletion of lipid A makes the bacteria susceptible to various classes of hydrophobic antibiotics and degradation of O side chains determine the smoothness and roughness of the LPS [66, 67]. The use of ethylenediaminetetraacetic acid (EDTA), some organic acids like lactic acid and citric acid are found to alter the impermeability of the *Pseudomonas* species. These chelating agents can neutralize the negatively charged oligosaccharide core by binding with the (Mg^{2+}) cations in the LPS molecule and promotes the removal of LPS molecules [68]. The accumulation of aminoglycoside level decreases in the case of *P. aeruginosa* leading to low uptake and hence shows impermeability resistance which has been reported in the strains isolated from the cystic fibrosis patients [58]. Similarly, tobramycin resistance due to impermeability was seen when studied for endocarditis in case of rabbits.

4.3 Through the efflux system

The drug efflux system in bacteria includes three major components i.e. outer membrane channel-forming protein (OMF), resistance nodulation division (RND) which helps in drug-protein antiport process and the membrane fusion protein that acts as a periplasmic link between above two components [69]. The *mexXY* operon codes the inner membrane protein (i.e. *MexY*) and periplasmic protein (i.e. *MexX*). Resistance nodulation division (RND) involves the *MexXY* efflux system which develops the resistance in *Pseudomonas* species [70, 71]. MexAB-OprM shows resistance against ticarcillin, broad-spectrum cephalosporin and β-lactam of clinical isolates, while the combination of MexAB-OprM, MexCD-OprJ and MexXY-OprM shows the carbapenem resistance [72]. The bacterial isolates like *Burkholderia pseudomallei* and *Escherichia coli* involve the three component systems known as RND type aminoglycoside efflux system. Treatment with ofloxacin and gentamicin increases the level of *MexXY* expression in case of mutants compared to wild-type strains [73, 74]. The wild-type of strains of *Pseudomonas* is resistance to the antibiotic classes like tetracyclines, aminoglycosides, glycylcyclines and erythromycin but the *MexXY* can express in presence of diverse class of antibiotics like lincomycin [75], macrolides [76], fluoroquinolones [77], chloramphenicol [30], β-lactams [72], novobiocin [78] along with the wild type of antibiotic classes. In the reduced aminoglycosides condition both adaptive and impermeability resistance in the *Pseudomonas* sp. is expressed. The expression of *MexXY* gene is regulated by *mexZ* repressor, present in the upstream region of *MexXY* region of the gene and belongs to tetracycline repressor protein (TetR) and AcrR repressor protein family [79].

4.4 Modification in the outer membrane

The exoskeleton of the Gram-negative bacteria is present to resist against the adverse environmental conditions. Likewise, the outer membrane of *P. aeruginosa* is designed in such a way that it can permit small hydrophilic molecules and inhibit larger molecules such as antibiotics [80]. Due to the crucial arrangement of aquaporin proteins in the cell membrane, the small hydrophilic antibiotics of quinolone and β-lactam classes can pass through the outer membrane. *P. aeruginosa* strains produce four major aquaporins (i.e. oprP, oprD, oprF and oprB) and two minor aquaporins (i.e. oprC, oprE) whereas the mutant strains lack oprF [81, 82]. The oprD is a specialized porin molecule present in bacterial membrane that helps in the process of up-taking positively charged amino acids like arginine and lysine [83]. The minimum inhibitory concentration increases due to the loss of oprD porin from the outer membrane of the *Pseudomonas* sp. thus increasing the resistance to imipenem class of antibiotics [84]. As the porin channels are impermeable to the polymyxin E and aminoglycoside, these molecules bind with the LPS present in the outer membrane, destructs the barrier and allows the antibiotics to enter into the bacterial cells [85]. Through this mechanism the aminoglycosides can enter into the cytoplasm of the bacterial cell and disturb the protein synthesis process in the ribosomes that kills the bacteria simultaneously. But the overexpression of the oprH an outer membrane protein [86], prevents the binding of antibiotics to LPS making it resistant for laboratory strains of *Pseudomonas* species.

4.5 Resistance by biofilm

Bacterial communities aggregate themselves to a substratum and encapsulated in a proteinous polysaccharide of matrix evolved during adverse environmental condition such as various irradiation treatments and therapy which is known as biofilm. Mostly these polysaccharide/polymeric matrix leads to the formation of biofilms over a water surface and shows resistance and enhances their survivability against the antimicrobial agents [87, 88]. The formation of biofilm is predominantly found in case of various biomedical instruments such as catheter, implants, ventilator and dialyser used patients residing in the hospital [89]. The bacteria are found to evade from host immune response due to the formation of biofilms and helps in promoting collateral damage to the tissues. Only few antibiotic classes act as an effective bactericidal agent for the free-floating bacteria but it fails to act against the bacteria forming biofilms as the biofilms are 1000 times more invulnerable to it [90, 91]. During environmental stress conditions, the bacteria change from free-living unicellular form to the planktonic form and then to the attached biofilm structure which enables the survivability of the bacteria. The matured biofilm starts to segregate from a place and develop an immobile structure in the new surfaces for colonization [92, 93]. The chemical therapy of antibiotics was not effective as the molecules cannot penetrate into the complex biofilm matrix due to the production of cover like exopolysaccharides matrix known as glycocalyx [94, 95]. Mostly the pathovars of *P. aeruginosa* forms the biofilm in the dialysis membrane and restricts the diffusion of piperacillin antibiotic into the complex aggregation [96]. It is pertinent to mention here that the bacterial biofilm is resistant to various classes/groups of antibiotics.

4.6 Resistance by quorum sensing

The *P. aeruginosa* has been found to be resistive to various bactericidal agents and mainly infects to the people suffering from HIV-AIDS and cancer due to the compromised immune system, use of broad-spectrum antibiotics for a longer duration and dependency on life support medical devices like a catheter, ventilator and dialyser. The bacteria communicate with each other by secreting extracellular signaling molecules known as autoinducer. The autoinducer level is directly proportional to the growth of bacterial population, hence with the increase in bacterial population the accumulation of autoinducer in the environment is at the peak [10, 97]. This process of production, release of signaling molecules is termed as quorum sensing.

There are four types of quorum sensing pathways discovered for the *P. aeruginosa* species which includes the LasR and LasI, RhlR and RhlI, PqsR-quinolone controlled system and the integrated quorum sensing (IQS) system which works under limiting conditions of phosphate [98, 99]. The formation of complexes of LasR with 3-oxo-C12-HSL activates the LasI synthase gene which helps in the process of autoinduction. The LasR complex regulates the expression of rhlI and rhlR genes along with the PQS systems which are related to the second and third mode of quorum sensing system of pathway respectively. The activation of its own regulon by the binding of C4-HSL with RhlR induces the second induction processes. The activation of RhlR is induced by PqsR-PQS complex which regulates the three modes of signaling in Quorum sensing along with inhibits the expression of the pqsR and pqsABCD. The ratio of 3-oxo-C12-HSL to C4-HSL gives an idea about the activation of PQS [100, 101]. The virulence property of *P. aeruginosa* is controlled by the RhlR along with C4-HSL and PqsR or LasR. Incase of the isolates of *P. aeruginosa* from the cystic fibrosis patients the mutations in the LasR supplies the autoinducer as there is the necessity of phosphate starvation protein (PhoB). This LasR activates the expression of pqs genes by the production of IQS which expresses the rhl gene hence shows the pathogenicity [102].

4.7 Others

Pseudomonas species also include the resistance mechanism like adaptive resistance, acquired resistance and intrinsic resistance which further helps in the increasing the resistivity of the pathogen to a wide range of antibiotic class.

4.7.1 Adaptive resistance

The resistance which is dependent on the physical and chemical stresses, growth states and promotes the initiation of the regular processes inside the cell in the presence of antibiotics and reverts back to the primary condition in the removal of the inducers are known as adaptive resistance [103, 104]. Previous research studies manifested that the resistance is due to many factors like the use of sub-inhibitory concentration of antimicrobial agents, polyamines, heat shock, SOS response, pH imbalance and anaerobiosis condition [105, 106]. *P. aeruginosa* was found to develop adaptive resistance against divalent Ca^{2+} and Mg^{2+} ions and the polymyxins which are controlled by *PmrAB* and *PhoPQ* pathways [107]. *P. aeruginosa* gradually reduces susceptibility in the presence of antibiotics and is altered in absentia this phenomena are reversible in nature and scientifically termed as the adaptive resistance [108]. The extensive studies revealed that adaptive resistance can also be developed in both *in vivo* and *in vitro* conditions due to the administration of antibiotics into the bacterial culture for few hours and this resistance disappears after the removal of antibiotics from the media [109]. But it is observed that the organism shows resistance when there is a low accumulation of the aminoglycosides. The resistance induced through drug efflux system and due to the gene expression associated with anaerobic respiration. The bacteria were grown in the anaerobic condition and nitrate environment to check the accumulation and the uptake of aminoglycoside and found that *P. aeruginosa* is capable of showing resistance in the anaerobic conditions [110].

4.7.2 Acquired resistance

The acquired resistance involves the transfer of plasmids, prophages, DNA elements and transposons by means of transduction, transformation and conjugation. This horizontal transfer shows the β-lactam and aminoglycoside resistance in *P. aeruginosa* [111]. The chemical modification of the aminoglycosides alters the affinity of a 30S subunit of ribose sugar to the target. Antibiotic drugs like cephalosporin, carbapenem [112] and penicillin [113] help in the process of development of resistance property in case to *P. aeruginosa* [112]. The mutational resistance occurred due to the formation of biofilms and the action of DNA-damaging agents. The mutation frequency is found to be increased by 10-fold, greater than 100-fold and 70-fold if the resistance is caused by meropenem [85], ciprofloxacin and if any mutation in genes respectively [114]. The downregulation of antioxidant enzymes damages the DNA in the biofilms. The library screening of cystic fibrosis (CF) patients describe that there are various mutators play a significant role during the early infection stages, mutL and mutS are the hypermutators which are widely found. The mutation in genes mexR, mexZ and nfxB is due to the overexpression of MexAB-OprM, MexXY-OprM and MexCD-OprJ efflux pump respectively. OprD is a porin that suppresses the uptake of imipenem [115] and another antibiotic [116] leading to the clinical resistance. The ampC β-lactamase, AmpD mutate and controls the activity of AmpR regulator [117]. The *P. aeruginosa* clinical strain shows resistance to mutations in gyrase (gyrA) and gyrB as well as parC and parE. Overlay we can demonstrate that the mutations in the unrelated genes give rise to acquired resistance against different antibiotics.

4.7.3 Intrinsic resistance

The intrinsic resistance is due to the combination of the efflux system along with the β-lactamase and the low outer membrane permeability, the entry of antibiotic molecules through the outer membrane of the bacteria [8]. The increase in antibiotic concentration in the environment helps in the low permeability of the outer membrane permits the entry of larger compounds and antibiotics into the cell with the help of porin protein channels and makes the bacteria resistant this slow process helps in increased resistance of the organism [83, 118]. The intrinsic resistance is carried out by the help of multi-drug efflux systems like MexAB-OprM and MexXY-OprM operon along with the inactivation of enzyme β-lactams by hydrolysis [119, 120].

5. Impact of *Pseudomonas aeruginosa* on the economy

The low membrane permeability, overexpression of efflux pump and deletion of porin channels are the cause behind the resistance of *Pseudomonas* species. *P. aeruginosa* was predominantly found in the ICUs of European continents hence put in the list of "ESKAPE" pathogens by the Infectious Disease Society of America [121, 122]. The existing antibacterial agents are not effective against these isolates and hence a severe threat for public health. A study in China for the bacterial resistance surveillance demonstrated that the resistance in case of hospital-acquired infection (HAI) is prevalence than community-acquired infection (CAI) [123]. Relatively few studies explained about the outbreak of Multi-drug resistance (MDR) in *P. aeruginosa* species. The worldwide study of *Pseudomonas* infections gives us the idea that in the year 2002 14% and in 2003 9.9% resistance were found in ICU isolates and nosocomial infections in United states [77]. During 1997–1999 8.2% and 4.7% of resistance were due to nosocomial infections in South America and Europe respectively [124, 125]. In 2001 2.8% and in 2005 6.9% of resistance were due to nosocomial infections in Japan [126] and Malaysia [127].

The National Nosocomial Infection Surveillance System (NNIS) also conducted the study for statistical analysis of the resistance developed by the hospital strains of *P. aeruginosa* and define that the hospital samples are more resistive to various groups of antibiotic classes [128]. The resistance to various classes of antibiotic by *P. aeruginosa* is a new threat to our defence system as once compromised it will be a difficult task to control the spread and infection of the bacteria among the living system. It has been also reported that the bacteraemia was not in control by the administration of antibiotics as it was spread by the antibiotic-resistant strains of *P. aeruginosa* [129].

Due to hospitalization for a significant period of time in the ICU [130] of a patient suffering from respiratory disorder [110], kidney disease [89] and other diseases which needs the ventilator along with the medical device installation are more prone to the infection of *P. aeruginosa* [131]. The administration of various drugs makes the *Pseudomonas* strain more resistive due to mechanisms like multi-drug-resistance (MDR), efflux systems, and loss of porin proteins from the outer membrane. Extensive research work is necessary to understand the infection mechanism and the development of resistance in the bacteria, the suitable combination of antibiotic molecules which will overcome the resistant behaviour and eradication of the bacterial biofilm without affecting the other processes in the living beings.

6. Mitigation of resistance

The eradication of the resistance is highly necessary for the prevention followed by cure to *Pseudomonas* infection for healthy sustenance. So, research is still going on to overcome the resistance by the organism and combinational therapeutic approach is found to be an effective tool against the resistance of the *Pseudomonas* species.

Cross-infection through hospital personnel gives rise to 30–40% of infection so irrespective of cost and time use of masks, cloths, gloves, antiseptics for the proper isolation can minimize the resistant developed in the pathovars [132]. It was observed that usual laboratory methods failed to detect the Antimicrobial-Drug resistance hence new testing methods, standards and guidelines implemented by various national and international clinical research groups for the early detection and control its outbreak [133]. The synergistic of two or more anti-bactericidal molecules is found to be an effective than monotherapy to overcome the resistance. The combination of polymixin with tobramycin is found to be an effective antimicrobial for inhibition in the formation of biofilms [134]. The combinational administration of tobramycin with aminoglycoside and macrolide clarithromycin shows a devastating effect against the biofilm [79]. Likewise, the integration of azithromycin with the tobramycin helped to destroy the bacterial biofilm when treated with *in vitro* condition [135].

The use of nitric oxide (NO) was reported to trigger the downstream of signal processing in quorum sensing and hence the production of cyclic-di-GMP decreases hence the extracellular matrix of biofilm get destroyed [136]. The introduction of deoxyribonuclease (DNAse) directly into the biofilm of the bacterial colony as it digests the environmental DNA (eDNA) enzymatically. The *P. aeruginosa* contains a molecule known as acyl-homoserine lactones (AHL), the blockage of signaling of this molecule prevents the formation of biofilms [137]. The *rsaL* gene expression acts as a negative regulator of the *lasI* gene expression which is responsible for the quorum sensing in the strains of *P. aeruginosa* [138]. The *PmrAB* and *PhoPQ* can alter the permeability of the outer membrane as the level of divalent ions decrease it increase the extracellular DNA in the biofilms and shows resistance to cationic bactericidal peptides and polymyxins [139]. Due to this phenomenon, the addition of amino arabinose to the 1st and 4th phosphate position in lipid A of the LPS and the net negative charge neutralized and the cations can enter into the bacterial cell [140].

The medical equipment and the biomaterial use for implantation purpose are coated with silver which reduces the adherence and biofilm producing ability of the bacteria. The novel compounds like curlicides and pilicides have been reported to inhibit the role of adhesin molecules and hence reduces the formation of biofilms on the surfaces. The use of nanomaterials of graphene and zinc as the coating of biomedical implants are found to be effective against the biofilm formation [141]. In some instances, it is necessary to replace the device after prolonged use with the patient/s. The small molecular artificially engineered peptide 1018 was discovered with the anti-biofilm activity [142].

The pharmaceutical industries are working towards the development of vaccines to tackle the antimicrobial resistance and few are under clinical trials which are believed to be effective against the resistance [143, 144]. There are several vaccines such as polysaccharide-protein conjugates, LPS-O antigen, OprI and OprF membrane protein, live-attenuated, flagella and DNA vaccines are known to be invented for the control of antimicrobial resistance of *P. aeruginosa*. But the recombinant vaccine IC43, OprI and OprF and flagella vaccines are found effective and are under clinical trials for cystic fibrosis patients [145]. Apart from the above various NGOs and educational groups are playing a great role to educate the students, doctors, hospital personnel and society by making people aware about the use of proper dose and medicines by consulting the physician along with the maintenance of hygiene in the surroundings.

7. Concluding remarks

P. aeruginosa as an emerging human pathogen causes an array of diseases in immunocompromised patients, newborns as well as healthy persons. The infection as a biofilm is much more severe than monoculture. Various antimicrobial/antibiotics treatment leads to not only increases the resistance in different strains of *P. aeruginosa* but also increase the disease incidence. The present chapter clearly enlightens various mechanisms of infection of *P. aeruginosa*, its biofilms and resistance pathways/mechanisms, global impact due to infections which further paves the way for various remediation in future through improved implementations of genetic engineering and advances nanotechnology tools.

Author details

Swaraj Mohanty, Bighneswar Baliyarsingh and Suraja Kumar Nayak*
Department of Biotechnology, College of Engineering and Technology (CET), Biju Patnaik University of Technology (BPUT), Bhubaneswar, Odisha, India

*Address all correspondence to: surajnayak3@gmail.com

References

[1] Kerckhoffs AP, Ben-Amor K, Samsom M, van der Rest ME, de Vogel J, Knol J, et al. Molecular analysis of faecal and duodenal samples reveals significantly higher prevalence and numbers of *Pseudomonas aeruginosa* in irritable bowel syndrome. Journal of Medical Microbiology. 2011;**60**(2):236-245

[2] Murga R, Miller J, Donlan R. Biofilm formation by gram-negative bacteria on central venous catheter connectors: Effect of conditioning films in a laboratory model. Journal of Clinical Microbiology. 2001;**39**(6):2294-2297

[3] Stickler D. Susceptibility of antibiotic-resistant Gram-negative bacteria to biocides: A perspective from the study of catheter biofilms. Journal of Applied Microbiology. 2002;**92**:163S-170S

[4] Johani K, Abualsaud D, Costa DM, Hu H, Whiteley G, Deva A, et al. Characterization of microbial community composition, antimicrobial resistance and biofilm on intensive care surfaces. Journal of Infection and Public Health. 2018;**11**(3):418-424

[5] De Silva B, Wimalasena S, Hossain S, Pathirana H, Heo G-J. Characterization of quinolone resistance of *Pseudomonas aeruginosa* isolated from pet chinese stripe-necked turtles (*Ocadia sinensis*). Asian Journal of Animal and Veterinary Advances. 2017;**12**(3):152-160

[6] Peleg AY, Hooper DC. Hospital-acquired infections due to gram-negative bacteria. New England Journal of Medicine. 2010;**362**(19):1804-1813

[7] Sriramulu DD, Lünsdorf H, Lam JS, Römling U. Microcolony formation: A novel biofilm model of *Pseudomonas aeruginosa* for the cystic fibrosis lung. Journal of Medical Microbiology. 2005;**54**(7):667-676

[8] Mesaros N, Nordmann P, Plésiat P, Roussel-Delvallez M, Van Eldere J, Glupczynski Y, et al. *Pseudomonas aeruginosa*: Resistance and therapeutic options at the turn of the new millennium. Clinical Microbiology and Infection. 2007;**13**(6):560-578

[9] Overhage J, Bains M, Brazas MD, Hancock RE. Swarming of *Pseudomonas aeruginosa* is a complex adaptation leading to increased production of virulence factors and antibiotic resistance. Journal of Bacteriology. 2008;**190**(8):2671-2679

[10] Smith RS, Iglewski BH. *P. aeruginosa* quorum-sensing systems and virulence. Current Opinion in Microbiology. 2003;**6**(1):56-60

[11] Tenover FC. Mechanisms of antimicrobial resistance in bacteria. American Journal of Infection Control. 2006;**119**(6):S3-S10

[12] Normark BH, Normark S. Evolution and spread of antibiotic resistance. Journal of Internal Medicine. 2002;**252**(2):91-106

[13] Tam VH, Rogers CA, Chang K-T, Weston JS, Caeiro J-P, Garey KW. Impact of multidrug-resistant *Pseudomonas aeruginosa* bacteremia on patient outcomes. Antimicrobial Agents and Chemotherapy. 2010;**54**(9):3717-3722

[14] Miko BA, Pereira MR, Safdar A. Respiratory tract infections: Sinusitis, bronchitis, and pneumonia. In: Principles and Practice of Transplant Infectious Diseases. New York, NY: Springer; 2019. pp. 339-349

[15] Moore LS, Cunningham J, Donaldson H. A clinical approach to managing *Pseudomonas aeruginosa* infections. British Journal of Hospital Medicine. 2016;77(4):C50-CC4

[16] Olszewski AE, Karandikar MV, Surana NK. Aeromonas as a cause of purulent folliculitis: A case report and review of the literature. Journal of the Pediatric Infectious Diseases Society. 2017;**6**(1):e1-e3

[17] Aliyu I, Kumurya A, Bala J, John O. Bacteriology of otitis media and its host-environmental-infection factors. Asia Pacific Environmental and Occupational Health Journal. 2017;**3**(1):20-27

[18] Heward E, Cullen M, Hobson J. Microbiology and antimicrobial susceptibility of otitis externa: A changing pattern of antimicrobial resistance. The Journal of Laryngology & Otology. 2018;**132**(4):314-317

[19] Saraswathi P, Beuerman RW. Corneal biofilms: From planktonic to microcolony formation in an experimental keratitis infection with *Pseudomonas aeruginosa*. The Ocular Surface. 2015;**13**(4):331-345

[20] Kugadas A, Christiansen SH, Sankaranarayanan S, Surana NK, Gauguet S, Kunz R, et al. Impact of microbiota on resistance to ocular *Pseudomonas aeruginosa*-induced keratitis. PLoS Pathogens. 2016;**12**(9):e1005855

[21] Cole SJ, Records AR, Orr MW, Linden SB, Lee VT. Catheter-associated urinary tract infection by *Pseudomonas aeruginosa* is mediated by exopolysaccharide independent biofilms. Infection and Immunity. 2014;**82**(5):2048-2058

[22] Giordano P, Song J, Pertel P, Herrington J, Kowalsky S. Sequential intravenous/oral moxifloxacin versus intravenous piperacillin-tazobactam followed by oral amoxicillin-clavulanate for the treatment of complicated skin and skin structure infection. International Journal of Antimicrobial Agents. 2005;**26**(5):357-365

[23] Smith DJ, Ramsay KA, Yerkovich ST, Reid DW, Wainwright CE, Grimwood K, et al. *Pseudomonas aeruginosa* antibiotic resistance in A ustralian cystic fibrosis centres. Respirology. 2016;**21**(2):329-337

[24] Falagas ME, Koletsi PK, Bliziotis IA. The diversity of definitions of multidrug-resistant (MDR) and pandrug-resistant (PDR) *Acinetobacter baumannii* and *Pseudomonas aeruginosa*. Journal of Medical Microbiology. 2006;**55**(12):1619-1629

[25] Torres C, Perlin MH, Baquero F, Lerner DL, Lerner SA. High-level amikacin resistance in *Pseudomonas aeruginosa* associated with a 3′-phosphotransferase with high affinity for amikacin. International Journal of Antimicrobial Agents. 2000;**15**(4):257-263

[26] Lin D, Foley S, Qi Y, Han J, Ji C, Li R, et al. Characterization of antimicrobial resistance of *Pseudomonas aeruginosa* isolated from canine infections. Journal of Applied Microbiology. 2012;**113**(1):16-23

[27] Cézard C, Farvacques N, Sonnet P. Chemistry and biology of pyoverdines, Pseudomonas primary siderophores. Current Medicinal Chemistry. 2015;**22**(2):165-186

[28] De Kievit TR, Parkins MD, Gillis RJ, Srikumar R, Ceri H, Poole K, et al. Multidrug efflux pumps: Expression patterns and contribution to antibiotic resistance in *Pseudomonas aeruginosa* biofilms. Antimicrobial Agents and Chemotherapy. 2001;**45**(6):1761-1770

[29] Imamovic L, Ellabaan MMH, Machado AMD, Citterio L, Wulff T, Molin S, et al. Drug-driven phenotypic convergence supports rational treatment strategies of chronic infections. Cell. 2018;**172**(1-2):121-134. e14

[30] Dean CR, Visalli MA, Projan SJ, Sum P-E, Bradford PA. Efflux-mediated

resistance to tigecycline (GAR-936) in *Pseudomonas aeruginosa* PAO1. Antimicrobial Agents and Chemotherapy. 2003;**47**(3):972-978

[31] Lomovskaya O, Warren MS, Lee A, Galazzo J, Fronko R, Lee M, et al. Identification and characterization of inhibitors of multidrug resistance efflux pumps in *Pseudomonas aeruginosa*: Novel agents for combination therapy. Antimicrobial Agents and Chemotherapy. 2001;**45**(1):105-116

[32] Piddock LJ. Clinically relevant chromosomally encoded multidrug resistance efflux pumps in bacteria. Clinical Microbiology Reviews. 2006;**19**(2):382-402

[33] Muller C, Plésiat P, Jeannot K. A two-component regulatory system interconnects resistance to polymyxins, aminoglycosides, fluoroquinolones, and β-lactams in *Pseudomonas aeruginosa*. Antimicrobial Agents and Chemotherapy. 2011;**55**(3):1211-1221

[34] Nouri R, Ahangarzadeh Rezaee M, Hasani A, Aghazadeh M, Asgharzadeh M. The role of gyrA and parC mutations in fluoroquinolones-resistant *Pseudomonas aeruginosa* isolates from Iran. Brazilian Journal of Microbiology. 2016;**47**(4):925-930

[35] Olaitan AO, Morand S, Rolain J-M. Mechanisms of polymyxin resistance: Acquired and intrinsic resistance in bacteria. Frontiers in Microbiology. 2014;**5**:643

[36] Moskowitz SM, Brannon MK, Dasgupta N, Pier M, Sgambati N, Miller AK, et al. PmrB mutations promote polymyxin resistance of *Pseudomonas aeruginosa* isolated from colistin-treated cystic fibrosis patients. Antimicrobial Agents and Chemotherapy. 2012;**56**(2):1019-1030

[37] Mawabo IK, Noumedem JA, Kuiate JR, Kuete V. Tetracycline improved the efficiency of other antimicrobials against gram-negative multidrug-resistant bacteria. Journal of Infection and Public Health. 2015;**8**(3):226-233

[38] Nejma MB, Sioud O, Mastouri M. Quinolone-resistant clinical strains of *Pseudomonas aeruginosa* isolated from University Hospital in Tunisia. 3 Biotech. 2018;**8**(1):1

[39] Jacoby GA. Plasmid-mediated quinolone resistance. In: Antimicrobial Drug Resistance. Berlin: Springer; 2017. pp. 265-268

[40] Barnes MD, Taracila MA, Rutter JD, Bethel CR, Galdadas I, Hujer AM, et al. Deciphering the evolution of cephalosporin resistance to ceftolozane-tazobactam in *Pseudomonas aeruginosa*. MBio. 2018;**9**(6):e02085-e02018

[41] Streeter K, Katouli M. *Pseudomonas aeruginosa*: A review of their pathogenesis and prevalence in clinical settings and the environment. Infection, Epidemiology and Microbiology. 2016;**2**(1):25-32

[42] Gellatly SL, Hancock RE. *Pseudomonas aeruginosa*: New insights into pathogenesis and host defenses. Pathogens and Disease. 2013;**67**(3):159-173

[43] Alhazmi A. Pseudomonas aeruginosa-pathogenesis and pathogenic mechanisms. International Journal of Biology. 2015;**7**(2):44

[44] De Bentzmann S, Plésiat P. The *Pseudomonas aeruginosa* opportunistic pathogen and human infections. Environmental Microbiology. 2011;**13**(7):1655-1665

[45] Kipnis E, Sawa T, Wiener-Kronish J. Targeting mechanisms of *Pseudomonas aeruginosa* pathogenesis. Médecine et Maladies Infectieuses. 2006;**36**(2):78-91

[46] Sousa A, Pereira M. *Pseudomonas aeruginosa* diversification during infection development in cystic fibrosis lungs—A review. Pathogens. 2014;**3**(3):680-703

[47] Azam MW, Khan AU. Updates on the pathogenicity status of *Pseudomonas aeruginosa*. Drug Discovery Today. 2019;**24**(1):350-359

[48] Feltman H, Schulert G, Khan S, Jain M, Peterson L, Hauser AR. Prevalence of type III secretion genes in clinical and environmental isolates of *Pseudomonas aeruginosa*. Microbiology. 2001;**147**(10):2659-2669

[49] Krueger KM, Barbieri JT. The family of bacterial ADP-ribosylating exotoxins. Clinical Microbiology Reviews. 1995;**8**(1):34-47

[50] Rüssmann H. Inverted pathogenicity: The use of pathogen-specific molecular mechanisms for prevention or therapy of disease. International Journal of Medical Microbiology. 2004;**293**(7-8):565-569

[51] Hazlett LD. Pathogenic mechanisms of *P. aeruginosa* keratitis: A review of the role of T cells, Langerhans cells, PMN, and cytokines. DNA and Cell Biology. 2002;**21**(5-6):383-390

[52] Moscoso JA, Mikkelsen H, Heeb S, Williams P, Filloux A. The *Pseudomonas aeruginosa* sensor RetS switches Type III and Type VI secretion via c-di-GMP signalling. Environmental Microbiology. 2011;**13**(12):3128-3138

[53] Sana TG, Hachani A, Bucior I, Soscia C, Garvis S, Termine E, et al. The second type VI secretion system of *Pseudomonas aeruginosa* strain PAO1 is regulated by quorum sensing and Fur and modulates internalization in epithelial cells. Journal of Biological Chemistry. 2012;**287**(32):27095-27105

[54] Kim JH, Park E-S, Shim JH, Kim M-N, Moon W-S, Chung K-H, et al. Antimicrobial activity of p-hydroxyphenyl acrylate derivatives. Journal of Agricultural and Food Chemistry. 2004;**52**(25):7480-7483

[55] Hobden JA. Pseudomonas aeruginosa proteases and corneal virulence. DNA and Cell Biology. 2002;**21**(5-6):391-396

[56] Wright GD. Bacterial resistance to antibiotics: Enzymatic degradation and modification. Advanced Drug Delivery Reviews. 2005;**57**(10):1451-1470

[57] Strateva T, Yordanov D. *Pseudomonas aeruginosa*—A phenomenon of bacterial resistance. Journal of Medical Microbiology. 2009;**58**(9):1133-1148

[58] Poole K. Aminoglycoside resistance in *Pseudomonas aeruginosa*. Antimicrobial Agents and Chemotherapy. 2005;**49**(2):479-487

[59] Bonomo RA, Szabo D. Mechanisms of multidrug resistance in Acinetobacter species and *Pseudomonas aeruginosa*. Clinical Infectious Diseases. 2006;**43**(Supplement_2):S49-S56

[60] Kotra LP, Haddad J, Mobashery S. Aminoglycosides: Perspectives on mechanisms of action and resistance and strategies to counter resistance. Antimicrobial Agents and Chemotherapy. 2000;**44**(12):3249-3256

[61] Marvig RL, Sommer LM, Molin S, Johansen HK. Convergent evolution and adaptation of *Pseudomonas aeruginosa* within patients with cystic fibrosis. Nature Genetics. 2015;**47**(1):57

[62] Azucena E, Mobashery S. Aminoglycoside-modifying enzymes: Mechanisms of catalytic processes and inhibition. Drug Resistance Updates. 2001;**4**(2):106-117

[63] Ramirez MS, Tolmasky ME. Aminoglycoside modifying enzymes.

Drug Resistance Updates. 2010;**13**(6):151-171

[64] Ruiz-Martínez L, López-Jiménez L, Fusté E, Vinuesa T, Martínez J, Viñas M. Class 1 integrons in environmental and clinical isolates of *Pseudomonas aeruginosa*. International Journal of Antimicrobial Agents. 2011;**38**(5):398-402

[65] Caroff M, Karibian D. Structure of bacterial lipopolysaccharides. Carbohydrate Research. 2003;**338**(23):2431-2447

[66] Crompton R, Williams H, Ansell D, Campbell L, Holden K, Cruickshank S, et al. Oestrogen promotes healing in a bacterial LPS model of delayed cutaneous wound repair. Laboratory Investigation. 2016;**96**(4):439-449

[67] Zhang L, Dhillon P, Yan H, Farmer S, Hancock RE. Interactions of bacterial cationic peptide antibiotics with outer and cytoplasmic membranes of *Pseudomonas aeruginosa*. Antimicrobial Agents and Chemotherapy. 2000;**44**(12):3317-3321

[68] Lambert P. Cellular impermeability and uptake of biocides and antibiotics in Gram-positive bacteria and mycobacteria. Journal of Applied Microbiology. 2002;**92**:46S-54S

[69] Stover C, Pham X, Erwin A, Mizoguchi S, Warrener P, Hickey M, et al. Complete genome sequence of *Pseudomonas aeruginosa* PAO1, an opportunistic pathogen. Nature. 2000;**406**(6799):959

[70] Dreier J, Ruggerone P. Interaction of antibacterial compounds with RND efflux pumps in *Pseudomonas aeruginosa*. Frontiers in Microbiology. 2015;**6**:660

[71] Morita Y, Nakashima K-I, Nishino K, Kotani K, Tomida J, Inoue M, et al. Berberine is a novel type efflux inhibitor which attenuates the MexXY-mediated aminoglycoside resistance in *Pseudomonas aeruginosa*. Frontiers in Microbiology. 2016;**7**:1223

[72] Villegas MV, Lolans K, Correa A, Kattan JN, Lopez JA, Quinn JP, et al. First identification of *Pseudomonas aeruginosa* isolates producing a KPC-type carbapenem-hydrolyzing β-lactamase. Antimicrobial Agents and Chemotherapy. 2007;**51**(4):1553-1555

[73] Jeannot K, Sobel ML, El Garch F, Poole K, Plésiat P. Induction of the MexXY efflux pump in *Pseudomonas aeruginosa* is dependent on drug-ribosome interaction. Journal of Bacteriology. 2005;**187**(15):5341-5346

[74] Chuanchuen R, Gaynor JB, Karkhoff-Schweizer R, Schweizer HP. Molecular characterization of MexL, the transcriptional repressor of the mexJK multidrug efflux operon in Pseudomonas aeruginosa. Antimicrobial Agents and Chemotherapy. 2005;**49**(5):1844-1851

[75] Livermore DM, Winstanley TG, Shannon KP. Interpretative reading: Recognizing the unusual and inferring resistance mechanisms from resistance phenotypes. Journal of Antimicrobial Chemotherapy. 2001;**48**(suppl_1):87-102

[76] Saiman L, Marshall BC, Mayer-Hamblett N, Burns JL, Quittner AL, Cibene DA, et al. Azithromycin in patients with cystic fibrosis chronically infected with *Pseudomonas aeruginosa*: A randomized controlled trial. Journal of the American Medical Association. 2003;**290**(13):1749-1756

[77] Livermore DM. Multiple mechanisms of antimicrobial resistance in *Pseudomonas aeruginosa*: Our worst nightmare? Clinical Infectious Diseases. 2002;**34**(5):634-640

[78] Zavascki AP, Carvalhaes CG, Picao RC, Gales AC. Multidrug-resistant *Pseudomonas aeruginosa* and

Acinetobacter baumannii: Resistance mechanisms and implications for therapy. Expert Review of Anti-Infective Therapy. 2010;**8**(1):71-93

[79] Poole K, Gilmour C, Farha MA, Parkins MD, Klinoski R, Brown ED. Meropenem potentiation of aminoglycoside activity against *Pseudomonas aeruginosa*: Involvement of the MexXY-OprM multidrug efflux system. Journal of Antimicrobial Chemotherapy. 2018;**73**(5):1247-1255

[80] Poole K. *Pseudomonas aeruginosa*: Resistance to the max. Frontiers in Microbiology. 2011;**2**:65

[81] Ochs MM, Bains M, Hancock RE. Role of putative loops 2 and 3 in imipenem passage through the specific porin OprD of *Pseudomonas aeruginosa*. Antimicrobial Agents and Chemotherapy. 2000;**44**(7):1983-1985

[82] Lee J-Y, Ko KS. OprD mutations and inactivation, expression of efflux pumps and AmpC, and metallo-β-lactamases in carbapenem-resistant *Pseudomonas aeruginosa* isolates from South Korea. International Journal of Antimicrobial Agents. 2012;**40**(2):168-172

[83] Hancock RE, Speert DP. Antibiotic resistance in *Pseudomonas aeruginosa*: Mechanisms and impact on treatment. Drug Resistance Updates. 2000;**3**(4):247-255

[84] Fernández L, Hancock RE. Adaptive and mutational resistance: Role of porins and efflux pumps in drug resistance. Clinical Microbiology Reviews. 2012;**25**(4):661-681

[85] Breidenstein EB, de la Fuente-Núñez C, Hancock RE. *Pseudomonas aeruginosa*: All roads lead to resistance. Trends in Microbiology. 2011;**19**(8):419-426

[86] Fernández L, McPhee JB, Tamber S, Brazas MD, Lewenza S, Hancock RE. Antibiotic resistance due to reduced uptake. In: Antimicrobial Drug Resistance. Cham: Springer; 2017. pp. 115-130

[87] Stewart PS, Costerton JW. Antibiotic resistance of bacteria in biofilms. The Lancet. 2001;**358**(9276):135-138

[88] Mah T-F, Pitts B, Pellock B, Walker GC, Stewart PS, O'toole GA. A genetic basis for *Pseudomonas aeruginosa* biofilm antibiotic resistance. Nature. 2003;**426**(6964):306

[89] Hall-Stoodley L, Costerton JW, Stoodley P. Bacterial biofilms: From the natural environment to infectious diseases. Nature Reviews. Microbiology. 2004;**2**(2):95

[90] Spoering AL, Lewis K. Biofilms and planktonic cells of *Pseudomonas aeruginosa* have similar resistance to killing by antimicrobials. Journal of Bacteriology. 2001;**183**(23):6746-6751

[91] Ma L, Conover M, Lu H, Parsek MR, Bayles K, Wozniak DJ. Assembly and development of the *Pseudomonas aeruginosa* biofilm matrix. PLoS Pathogens. 2009;**5**(3):e1000354

[92] Moreau-Marquis S, Stanton BA, O'Toole GA. *Pseudomonas aeruginosa* biofilm formation in the cystic fibrosis airway. Pulmonary Pharmacology & Therapeutics. 2008;**21**(4):595-599

[93] Murray TS, Egan M, Kazmierczak BI. *Pseudomonas aeruginosa* chronic colonization in cystic fibrosis patients. Current Opinion in Pediatrics. 2007;**19**(1):83-88

[94] Webb JS, Thompson LS, James S, Charlton T, Tolker-Nielsen T, Koch B, et al. Cell death in *Pseudomonas aeruginosa* biofilm development. Journal of Bacteriology. 2003;**185**(15):4585-4592

[95] Banin E, Vasil ML, Greenberg EP. Iron and *Pseudomonas aeruginosa* biofilm formation. Proceedings of the National Academy of Sciences. 2005;**102**(31):11076-11081

[96] Ryder C, Byrd M, Wozniak DJ. Role of polysaccharides in *Pseudomonas aeruginosa* biofilm development. Current Opinion in Microbiology. 2007;**10**(6):644-648

[97] Hentzer M, Wu H, Andersen JB, Riedel K, Rasmussen TB, Bagge N, et al. Attenuation of *Pseudomonas aeruginosa* virulence by quorum sensing inhibitors. The EMBO Journal. 2003;**22**(15):3803-3815

[98] Diggle SP, Winzer K, Chhabra SR, Worrall KE, Cámara M, Williams P. The *Pseudomonas aeruginosa* quinolone signal molecule overcomes the cell density-dependency of the quorum sensing hierarchy, regulates rhl-dependent genes at the onset of stationary phase and can be produced in the absence of LasR. Molecular Microbiology. 2003;**50**(1):29-43

[99] Dietrich LE, Price-Whelan A, Petersen A, Whiteley M, Newman DK. The phenazine pyocyanin is a terminal signalling factor in the quorum sensing network of *Pseudomonas aeruginosa*. Molecular Microbiology. 2006;**61**(5):1308-1321

[100] Parsek MR, Greenberg EP. Acyl-homoserine lactone quorum sensing in gram-negative bacteria: A signaling mechanism involved in associations with higher organisms. Proceedings of the National Academy of Sciences. 2000;**97**(16):8789-8793

[101] Rumbaugh KP, Griswold JA, Hamood AN. The role of quorum sensing in the in vivo virulence of *Pseudomonas aeruginosa*. Microbes and Infection. 2000;**2**(14):1721-1731

[102] Singh PK, Schaefer AL, Parsek MR, Moninger TO, Welsh MJ, Greenberg E. Quorum-sensing signals indicate that cystic fibrosis lungs are infected with bacterial biofilms. Nature. 2000;**407**(6805):762

[103] Oliver A, Cantón R, Campo P, Baquero F, Blázquez J. High frequency of hypermutable *Pseudomonas aeruginosa* in cystic fibrosis lung infection. Science. 2000;**288**(5469):1251-1253

[104] Hocquet D, Vogne C, El Garch F, Vejux A, Gotoh N, Lee A, et al. MexXY-OprM efflux pump is necessary for adaptive resistance of *Pseudomonas aeruginosa* to aminoglycosides. Antimicrobial Agents and Chemotherapy. 2003;**47**(4):1371-1375

[105] Skiada A, Markogiannakis A, Plachouras D, Daikos GL. Adaptive resistance to cationic compounds in *Pseudomonas aeruginosa*. International Journal of Antimicrobial Agents. 2011;**37**(3):187-193

[106] de la Fuente-Núñez C, Reffuveille F, Fernández L, Hancock RE. Bacterial biofilm development as a multicellular adaptation: Antibiotic resistance and new therapeutic strategies. Current Opinion in Microbiology. 2013;**16**(5):580-589

[107] McPhee JB, Lewenza S, Hancock RE. Cationic antimicrobial peptides activate a two-component regulatory system, PmrA-PmrB, that regulates resistance to polymyxin B and cationic antimicrobial peptides in *Pseudomonas aeruginosa*. Molecular Microbiology. 2003;**50**(1):205-217

[108] Fernández L, Gooderham WJ, Bains M, McPhee JB, Wiegand I, Hancock RE. Adaptive resistance to the "last hope" antibiotics polymyxin B and colistin in *Pseudomonas aeruginosa* is mediated by the novel two-component regulatory system

ParR-ParS. Antimicrobial Agents and Chemotherapy. 2010;**54**(8):3372-3382

[109] Sobel ML, McKay GA, Poole K. Contribution of the MexXY multidrug transporter to aminoglycoside resistance in *Pseudomonas aeruginosa* clinical isolates. Antimicrobial Agents and Chemotherapy. 2003;**47**(10):3202-3207

[110] Smith EE, Buckley DG, Wu Z, Saenphimmachak C, Hoffman LR, D'Argenio DA, et al. Genetic adaptation by *Pseudomonas aeruginosa* to the airways of cystic fibrosis patients. Proceedings of the National Academy of Sciences. 2006;**103**(22):8487-8492

[111] Magiorakos AP, Srinivasan A, Carey R, Carmeli Y, Falagas M, Giske C, et al. Multidrug-resistant, extensively drug-resistant and pandrug-resistant bacteria: An international expert proposal for interim standard definitions for acquired resistance. Clinical Microbiology and Infection. 2012;**18**(3):268-281

[112] Poirel L, Nordmann P. Carbapenem resistance in *Acinetobacter baumannii*: Mechanisms and epidemiology. Clinical Microbiology and Infection. 2006;**12**(9):826-836

[113] Levy SB, Marshall B. Antibacterial resistance worldwide: Causes, challenges and responses. Nature Medicine. 2004;**10**(12):S122-S129

[114] Breidenstein EBM. Global Regulation of the Lon Protease of *Pseudomonas aeruginosa* and Its Influence on Ciprofloxacin Resistance and Virulence [thesis]. Vancouver, BC V6T 1Z4, Canada: University of British Columbia; 2012

[115] Poirel L, Héritier C, Tolün V, Nordmann P. Emergence of oxacillinase-mediated resistance to imipenem in *Klebsiella pneumoniae*. Antimicrobial Agents and Chemotherapy. 2004;**48**(1):15-22

[116] Gibb AP, Tribuddharat C, Moore RA, Louie TJ, Krulicki W, Livermore DM, et al. Nosocomial outbreak of carbapenem-resistant *Pseudomonas aeruginosa* with a new blaIMP allele, blaIMP-7. Antimicrobial Agents and Chemotherapy. 2002;**46**(1):255-258

[117] Potron A, Poirel L, Nordmann P. Emerging broad-spectrum resistance in *Pseudomonas aeruginosa* and *Acinetobacter baumannii*: Mechanisms and epidemiology. International Journal of Antimicrobial Agents. 2015;**45**(6):568-585

[118] Li X-Z, Nikaido H. Efflux-mediated drug resistance in bacteria. Drugs. 2004;**64**(2):159-204

[119] Li X-Z, Zhang L, Poole K. Interplay between the MexA-MexB-OprM multidrug efflux system and the outer membrane barrier in the multiple antibiotic resistance of *Pseudomonas aeruginosa*. Journal of Antimicrobial Chemotherapy. 2000;**45**(4):433-436

[120] Masuda N, Sakagawa E, Ohya S, Gotoh N, Tsujimoto H, Nishino T. Substrate specificities of MexAB-OprM, MexCD-OprJ, and MexXY-oprM efflux pumps in *Pseudomonas aeruginosa*. Antimicrobial Agents and Chemotherapy. 2000;**44**(12):3322-3327

[121] Boucher HW, Talbot GH, Bradley JS, Edwards JE, Gilbert D, Rice LB, et al. Bad bugs, no drugs: No ESKAPE! An update from the Infectious Diseases Society of America. Clinical Infectious Diseases. 2009;**48**(1):1-12

[122] Vincent J-L, Rello J, Marshall J, Silva E, Anzueto A, Martin CD, et al. International study of the prevalence and outcomes of infection in intensive care units. Journal of the American Medical Association. 2009;**302**(21):2323-2329

[123] Xiao Y-H, Giske CG, Wei Z-Q, Shen P, Heddini A, Li L-J. Epidemiology and characteristics of antimicrobial resistance in China. Drug Resistance Updates. 2011;**14**(4-5):236-250

[124] Crespo M, Woodford N, Sinclair A, Kaufmann M, Turton J, Glover J, et al. Outbreak of carbapenem-resistant *Pseudomonas aeruginosa* producing VIM-8, a novel metallo-β-lactamase, in a tertiary care center in Cali, Colombia. Journal of Clinical Microbiology. 2004;**42**(11):5094-5101

[125] Doring G, Conway S, Heijerman H, Hodson M, Hoiby N, Smyth A, et al. Antibiotic therapy against *Pseudomonas aeruginosa* in cystic fibrosis: A European consensus. European Respiratory Journal. 2000;**16**(4):749-767

[126] Sekiguchi J-I, Asagi T, Miyoshi-Akiyama T, Kasai A, Mizuguchi Y, Araake M, et al. Outbreaks of multidrug-resistant *Pseudomonas aeruginosa* in community hospitals in Japan. Journal of Clinical Microbiology. 2007;**45**(3):979-989

[127] Hughes AJ, Ariffin N, Huat TL, Molok HA, Hashim S, Sarijo J, et al. Prevalence of nosocomial infection and antibiotic use at a university medical center in Malaysia. Infection Control and Hospital Epidemiology. 2005;**26**(1):100-104

[128] Rosenthal VD, Bijie H, Maki DG, Mehta Y, Apisarnthanarak A, Medeiros EA, et al. International Nosocomial Infection Control Consortium (INICC) report, data summary of 36 countries, for 2004-2009. American Journal of Infection Control. 2012;**40**(5):396-407

[129] Buehrle DJ, Shields RK, Clarke LG, Potoski BA, Clancy CJ, Nguyen MH. Carbapenem-resistant *Pseudomonas aeruginosa* bacteremia: Risk factors for mortality and microbiologic treatment failure. Antimicrobial Agents and Chemotherapy. 2017;**61**(1):e01243-e01216

[130] Trautmann M, Lepper PM, Haller M. Ecology of *Pseudomonas aeruginosa* in the intensive care unit and the evolving role of water outlets as a reservoir of the organism. American Journal of Infection Control. 2005;**33**(5):S41-S49

[131] Hauser AR, Cobb E, Bodí M, Mariscal D, Vallés J, Engel JN, et al. Type III protein secretion is associated with poor clinical outcomes in patients with ventilator-associated pneumonia caused by Pseudomonas aeruginosa. Critical Care Medicine. 2002;**30**(3):521-528

[132] Rahal JJ, Urban C, Segal-Maurer S. Nosocomial antibiotic resistance in multiple Gram-negative species: Experience at one hospital with squeezing the resistance balloon at multiple sites. Clinical Infectious Diseases. 2002;**34**(4):499-503

[133] Reis AO, Cordeiro JC, Machado AM, Sader HS. In vitro antimicrobial activity of linezolid tested against vancomycin-resistant enterococci isolated in Brazilian hospitals. Brazilian Journal of Infectious Diseases. 2001;**5**(5):243-251

[134] Zhanel GG, Mayer M, Laing N, Adam HJ. Mutant prevention concentrations of levofloxacin alone and in combination with azithromycin, ceftazidime, colistin (Polymyxin E), meropenem, piperacillin-tazobactam, and tobramycin against Pseudomonas aeruginosa. Antimicrobial Agents and Chemotherapy. 2006;**50**(6):2228-2230

[135] Nichols DP, Happoldt CL, Bratcher PE, Caceres SM, Chmiel JF, Malcolm KC, et al. Impact of azithromycin on the clinical and antimicrobial effectiveness of tobramycin in the treatment of cystic

fibrosis. Journal of Cystic Fibrosis. 2017;**16**(3):358-366

[136] Barraud N, Hassett DJ, Hwang S-H, Rice SA, Kjelleberg S, Webb JS. Involvement of nitric oxide in biofilm dispersal of *Pseudomonas aeruginosa*. Journal of Bacteriology. 2006;**188**(21):7344-7353

[137] Matsuo Y, Eda S, Gotoh N, Yoshihara E, Nakae T. MexZ-mediated regulation of mexXY multidrug efflux pump expression in *Pseudomonas aeruginosa* by binding on the mexZ-mexX intergenic DNA. FEMS Microbiology Letters. 2004;**238**(1):23-28

[138] Schuster M, Lostroh CP, Ogi T, Greenberg EP. Identification, timing, and signal specificity of *Pseudomonas aeruginosa* quorum-controlled genes: A transcriptome analysis. Journal of Bacteriology. 2003;**185**(7):2066-2079

[139] Nuri R, Shprung T, Shai Y. Defensive remodeling: How bacterial surface properties and biofilm formation promote resistance to antimicrobial peptides. Biochimica et Biophysica Acta (BBA)-Biomembranes. 2015;**1848**(11):3089-3100

[140] Schurck KN, Sampaio JL, Kiffer CR, Sinto S, Mendes CM, Hancock RE. Involvement of pmrAB and phoPQ in polymyxin B adaptation and inducible resistance in non-cystic fibrosis clinical isolates of Pseudomonas aeruginosa. Antimicrobial Agents and Chemotherapy. 2009;**53**(10):4345-4351

[141] Priyadarsini S, Mohanty S, Mukherjee S, Basu S, Mishra M. Graphene and graphene oxide as nanomaterials for medicine and biology application. Journal of Nanostructure in Chemistry. 2018;**8**(2):123-137

[142] Taylor PK, Yeung AT, Hancock RE. Antibiotic resistance in *Pseudomonas aeruginosa* biofilms: Towards the development of novel anti-biofilm therapies. Journal of Biotechnology. 2014;**191**:121-130

[143] Jansen KU, Anderson AS. The role of vaccines in fighting antimicrobial resistance (AMR). Human Vaccines & Immunotherapeutics. 2018;**14**(9):2142-2149

[144] Merakou C, Schaefers MM, Priebe GP. Progress toward the elusive *Pseudomonas aeruginosa* vaccine. Surgical Infections. 2018;**19**(8):757-768

[145] Pang Z, Raudonis R, Glick BR, Lin T-J, Cheng Z. Antibiotic resistance in *Pseudomonas aeruginosa*: Mechanisms and alternative therapeutic strategies. Biotechnology Advances. 2019;**37**(1):177-192

Section 2

Control Strategies

Chapter 4

Plant-Associated Microorganisms as a Potent Bio-Factory of Active Molecules against Multiresistant Pathogens

Felipe de Paula Nogueira Cruz, Andréa Cristina Bogas and Cristina Paiva de Sousa

Abstract

Antibiotic-resistant pathogens are a public health threat that has rapidly spread over decades due to continuous and uncontrolled administration of antimicrobial medicines, becoming an ever-increasing worldwide concern. Since the past decade, no significant innovations have been made, so the search for new compounds that face multidrug-resistant pathogens is critically important. Plant-symbiont microorganisms are capable of producing a variety of bioactive natural products, making it possible to treat several infectious diseases. Biotechnological processes using microorganisms have been increasing in recent years since the discovery of Paclitaxel, an important antimitotic produced by the endophyte *Taxomyces andreanae*. It was isolated for the first time from the native tree of Pacific *Taxus brevifolia*. Several studies have demonstrated the isolation and characterization of promising and potent substances capable of inhibiting these pathogens. In addition, both rhizospheric and endophytic communities represent an unexplored reserve of unique chemical structures for drug development. This chapter focuses on the potential of plant-derived microorganisms as a source of bioactive substances and the perspectives for further studies and their application.

Keywords: antimicrobial resistance, endophytes, natural products, rhizosphere, superbugs, *Streptomyces* spp.

1. Introduction

The discovery of medicines in the treatment of infectious diseases represents one of the most significant accomplishments of humankind. The introduction of antibiotics made it possible to treat previously incurable diseases.

Major classes of antibiotics were discovered between the 1940s and 1960s, where soil-derived actinobacteria produced most of them. However, several decades passed without significant innovations until the discovery and development of oxazolidinones in 2010 (**Figure 1**). Moreover, the continuous uncontrolled use of these medicines favored the rapid spread of resistant pathogens, where new compounds were discovered, and their introduction into clinical practice was not fast enough [1–5].

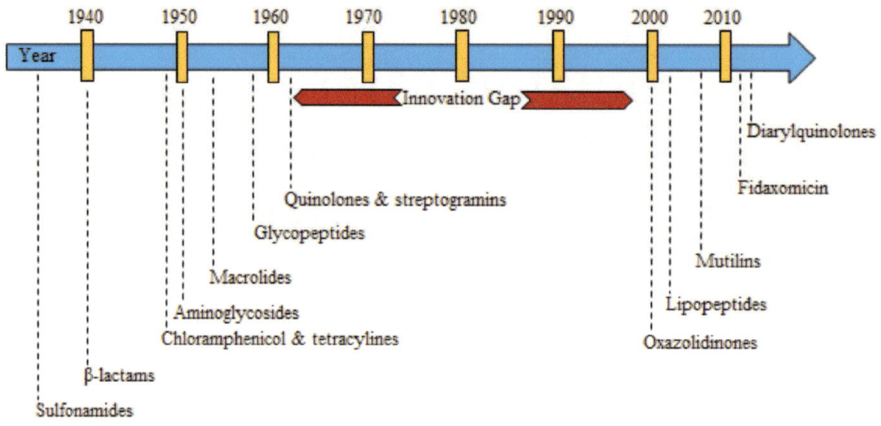

Figure 1.
Timeline of antibiotic discovery that shows no new classes of antibiotics between the years 1962 and 2000 adapted from: [6, 7].

The CDC (Centers for Disease Control and Prevention) has recognized the emerging antibiotic resistance as a significant threat to public health [8]. Superbugs, such as Methicillin-resistant *Staphylococcus aureus* (MRSA), show antibiotic resistance rates that surpass 50% in 5 out of 6 world regions; in contrast, the multidrug-resistant *Acinetobacter baumannii*, described as a dangerous agent by the Society of Infectious Diseases of America (SIDA), is a notable threat in intensive care units (ICUs) due to the development of resistance to broad-spectrum antibiotics [5, 8–10].

Therefore, the search for compounds and the exploration of niches that harbor microorganisms that produce bioactive metabolites are critically important [11–13]. Several studies have shown that plant tissues represent a rich source of natural products for pharmaceutical and biotechnological interest. Most of these compounds are produced by microorganisms that live in intimate interaction with the host plant without causing damage; therefore, they are known as endophytes [11, 14, 15].

In the same context, the rhizosphere's microbiome can exert profound direct and indirect effects on plant growth, nutrition, and health in natural ecosystems. Its micro-community (bacteria, oomycetes, viruses, archeas, fungi and arbuscular mycorrhizae) is attracted and fed by nutrients, exudates, border cells and mucilage that are released by the root of the plant [16].

Relevant studies have reported potent antimicrobial compounds, such as teixobactin, isolated from the non-cultivable bacterium *Eleftheria terrae* [17]. According to the authors in [17], teixobactin inhibits cell wall synthesis by binding to the highly conserved region of lipid precursors of peptidoglycan and teichoic acid. In addition, *S. aureus* and *Mycobacterium tuberculosis* did not develop resistance to teixobactin.

In the study by [18], endophytic fungi were isolated from the medicinal plant *Orthosiphon stamineus*, where 92% of them exhibited significant inhibitory activity against different species of bacterial pathogens and filamentous fungi.

Paenibacillus polymyxa can be found in several habitats. Its characteristic metabolism and production of substances enhance biotechnological applications based on the production of bioactive molecules. It is also widely applied in commercial agriculture as a bio-fertilizer grow plant promoter, biological control, and environmental remediation. In [19], *P. polymyxa* was endophytically isolated from *Prunus* spp., and the author reported the isolation of molecules which potently inhibited *S. aureus* and *E. coli*.

Herein, we address a review topic concerning the potential of rhizospheric and endophytic microorganisms as producers of antimicrobial compounds.

2. Endophytes: an overview

In 1866, de Bary outlined the first distinction between endophytes and plant pathogens. These microorganisms (typically fungi or bacteria) colonize the plant's internal tissues and live part of its life or its entire life cycle without causing apparent damage, establishing a mutualistic interaction with the host plant. Moreover, endophytes are capable of producing beneficial substances, such as alkaloids, enzymes, antibiotics and other compounds that protect and help the plant under stress conditions in exchange for nutrients and protection provided by the host plant [14, 15, 20–22].

In this context, plants have served humanity for centuries and led to the discovery of novel bioactive compounds. However, concerns regarding biodiversity and conservation, as well as large quantities of plant tissue, are required to produce sufficient yields of compounds [23]. According to [24], paclitaxel isolation requires about 10,000 kg of *T. brevifolia* bark to yield 1 kg. On the other hand, several studies have shown that endophytes may produce similar or even the same bioactive compounds as their plant hosts [20, 23, 25].

Fungi are skilled producers of natural products, including antitumor agents, cholesterol-lowering agents, immunosuppressants and antibiotics [25, 26]. The study by [27] detected potent antimicrobial properties of the natural product extract (NPE) of endophytic fungi associated with *Myrciaria floribunda*, *Alchornea castaneifolia* and *Eugenia aff. Bimarginata* against several pathogens. The methanolic extracts presented MIC values ranging from 7.8 to 1000 μg/mL against *C. krusei*, *C. parapsilosis*, *C. neoformans*, *C. albicans*, and *C. glabrata*. The inhibition of *S. aureus* and *B. cereus* ranged from 7.8 to >1000 μg/mL. Also, endophytic fungi were isolated from *Cinnamomum mercadoi*, a medicinal tree endemic to the Philippines. The ethyl acetate extract of *Fusarium* sp. presented moderate inhibition against *E. coli*, *E. aerogenes*, *S. aureus*, and *B. cereus* with minimum inhibitory concentrations of 2.1, 4.2, 4.2, and 3.8 mg/mL, respectively [28].

Therefore, the emerging use of endophytes in the research and development of new drugs represents the most successful example of bioactive natural products in medicine, pharmaceutical and biotechnological applications. **Table 1** provides an idea of some secondary metabolites of endophytic fungi and bacteria tested against resistant and multidrug-resistant microorganisms.

3. Rhizospheric microorganisms: an overview

The term rhizosphere was first used in 1904 by agronomist and plant physiologist Lorenz Hiltner to describe the interface between plant roots and the soil inhabited by a unique microbial community, which is influenced by the chemical release from plant roots [49]. In recent years, based on the relative proximity and influence to the root, the rhizosphere definition has been refined to include three zones: (i) endorhizosphere, which includes portions of the cortex and endoderm, where microorganisms and mineral ions occupy free space between cells (apoplastic space); (ii) rhizoplane, a middle zone adjacent to the root's epidermal cells and mucilage; and (iii) ectorhizosphere, which extends from the rhizoplane out into the bulk soil and is colonized by the microorganisms that are either free-living or non-symbionts [50, 51].

Endophytic fungi				
Endophyte	**Host plant**	**Compound**	**Target strain**	**Reference**
Trichoderma ovalisporum	*Panax notoginseng*	Shikimic acid	*S. aureus*	[29]
			E. coli	
Fusarium oxysporum	*Cinnamomum kanehirae*	Beauvericin	MR *S. aureus*	[30]
			B. subtilis (ATCC66333)	
Diaporthe phaseolorum	*Laguncularia racemosa*	3-Hidroxypropionic acid	*S. aureus*	[31]
			S. typhi	
Pestalotiopsis mangiferae	*Mangifera indica*	4-(2,4,7-trioxa-bicyclo[4.1.0]heptan-3-yl) phenol (1)	*B. subtilis* (MTCC 441)	[32]
			E. coli (MTCC 443)	
			P. aeruginosa (MTCC 424)	
			K. pneumonia (MTCC 109)	
			C. albicans (MTCC 227)	
Xylaria sp.	*Anoectochilus setaceus*	Helvolic acid	*B. subtilis* (UBC 344)	[33]
			MR *S. aureus* ATCC 33591	
Aspergillus terreus	*Carthamus lanatus*	(22E,24R)-stigmasta-5,7,22-trien-3-β-ol; Aspernolide F	*S. aureus* MRSA (ATCC 33591)	[34]
			C. neoformans (ATCC 90113)	
Hypocrea virens	*Premna serratifolia* L.	Gliotoxin	*C. neoformans* (ATCC 90113)	[35]
			B. subtilis (UBC 344)	
			S. aureus (ATCC 43300)	
			S. aureus MRSA (ATCC 33591)	
			E. coli (UBC 8161)	
			P. aeruginosa (ATCC 27853)	
			C. albicans (ATCC 90028)	
Aspergillus sp. TJ23	*Hypericum perforatum*	Spiroaspertrione A	*S. aureus* MRSA	[36]
Aspergillus sp. TJ23	*Hypericum perforatum*	Aspermerodione	*S. aureus* MRSA (ATCC 43300)	[37]

Endophytic fungi				
Endophyte	**Host plant**	**Compound**	**Target strain**	**Reference**
Phomopsis asparagi	*Paris polyphylla*	Diphenyl ethers derivates	*S. aureus* MRSA (ZR11)	[38]
Athelia rolfsii	*Coleus amboinicus* Lour.	Hemiterpenoid compounds	*S. aureus* (ATCC 25923)	[39]
			E. coli (ATCC 11229)	
			P. aeruginosa (ATCC 27853)	
			B. subtilis (ATCC 6633)	
			S. typhi (clinical)	
			S. mutans (ATCC 25175)	
Endophytic bacteria				
Streptomyces sp.	*Kandelia candel*	Indolosesquiterpenes	*S. aureus* MRSA	[40]
			Enterococcus faecalis VRE	
Streptomyces sp.	*Kandelia candel*	Eudesmene-type sesquiterpenes (kandenols)	*B. subtilis* (ATCC 6633)	[41]
S. sundarbansensis	*Fucus* sp.	Polyketides (2-hydroxy-5-((6-hydroxy-4-oxo-4H-pyran-2-yl) methyl) -2- propylchroman-4 one)	*S. aureus* MRSA (ATCC 43300)	[42]
Streptomyces sp.	*Dysophylla stellata*	2-amino-3,4-dihydroxy-5-methoxybenzamide	*E. coli*	[43]
			C. albicans	
Streptomyces sp.	*Dracaena cochinchinensis*	(*Z*)-tridec-7-ene-1,2,13-tricarboxylic acid	*S. epidermis* MRSA (ATCC 35984)	[44]
			S. aureus MRSA (ATCC 25923)	
		Actinomycin-D	*E. coli* (ATCC 25922)	
			K. pneumoniae (ATCC 13883)	
Streptomyces sp.	*Zingiber spectabile*	Diketopiperazine *cyclo* (tryptophanyl-prolyl); chloramphenicol	*S. aureus* MRSA (ATCC 43300)	[45]
			S. aureus MRSA (ATCC 49476)	
			S.aureus MRSA (ATCC 33591)	
Microbispora sp.	*Vochysia divergens*	1-Acetyl-β-carboline	*S. aureus* MSSA	[46]
			S. aureus MRSA	

Endophytic fungi				
Endophyte	**Host plant**	**Compound**	**Target strain**	**Reference**
S. cavourensis	*Cinnamomum cassia*	1-Monolinolein, bafilomycin D; nonactic acid; daidzein	*S. aureus* MRSA (ATCC 33591)	[47]
		3′-Hydroxydaidzein	*S. epidermidis* MRSE (ATCC 35984)	
Luteibacter sp.	*Astrocaryum sciophilum*	(*R*)-2-hydroxy-13 methyltetradecanoic acid, (*R*)-3-hydroxy-14methylpentadecanoic acid, (*S*)-β-hydroxypalmitic acid; (*R*)-3-hydroxy-15 methylhexadecanoic acid, (*R*)-3-hydroxy-13-methyltetradecanoic acid, 13-methyltetradecanoic acid; 9Z-hexadecenoic acid, 15-methyl-9*Z*-hexadecenoic acid	*S. aureus* MRSA	[48]
Streptomyces sp.	*Epipremnum aureum*	Phenylalanine-arginine β-naphthylamide	*Mycobacterium tuberculosis*	[49]
			B. cereus (ATCC11778)	
			E. faecium (ATCC51559)	
			A. baumannii (ATCC19606)	

Table 1.
Secondary metabolites produced by endophytic fungi and bacteria with antimicrobial activity (2010–2020).

The rhizosphere is a complex and dynamic region, where bacteria (including Plant Growth-Promoting Rhizobacteria—PGPR), fungi (including Arbuscular Mycorrhizal Fungi – AMF), oomycetes, viruses and archaea are attracted by chemical compounds (sugars, proteins, fatty acids, organics acids, vitamins, and other cellular components) released in the vicinity of the plant roots [16, 52, 53]. These rhizodeposits are used as carbon sources by microorganisms and represent an essential source of carbon allocated to the roots and available to plants through photosynthesis [54].

Rhizodeposits also contain secondary metabolites (flavonoids, antimicrobials and others) involved in establishing symbiosis or repelling plant pathogens and pests [55, 56].

The establishment of the symbiotic plant-PGPR interaction in the rhizosphere can favor the plant growth through direct and indirect mechanisms. The first one includes the fixation of atmospheric nitrogen [57], phosphate solubilization [58] or any other process capable of supplying the plant with some of its previously unavailable nutrients. Many PGPRs also produce phytohormones, such as auxins (Indole-3-acetic acid) and cytokinin, which exert strong effects on root and shoot growth, respectively [59–61]. The indirect mechanisms of plant growth prevent the deleterious effects of pathogens and include competition for nutrients and niches, induction of systemic resistance (Jasmonic acid (JA), and ethylene), and lytic enzymes (chitinase, pectinase, cellulase, glucanase, protease, xylanase), siderophore, bacteriocins and antibiotics production [62] (**Figure 2**).

The phyla of PGPR commonly found in the rhizosphere are Actinobacteria, Firmicutes, Proteobacteria and Bacteroidetes; among the main genera, *Burkholderia*, *Azotobacter, Pseudomonas*, *Bacillus, Methylobacterium*, *Serratia*, *Streptomyces*, *Azospirillum*, *Herbaspirillum* and *Rhizobium* can be mentioned [63, 64]. The latter can establish an effective symbiotic relationship with plant species of the Leguminosae

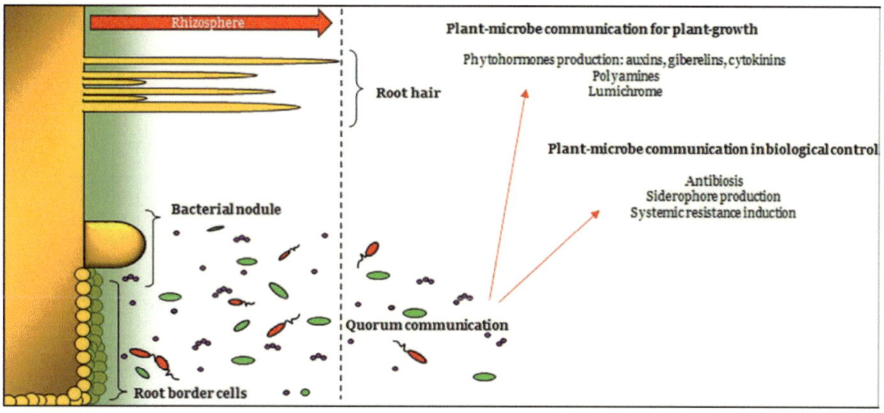

Figure 2.
Basic scheme of the rhizospheric space showing saprophytic and symbiotic bacteria and fungi, including arbuscular mycorrhizal fungi. Adapted from [16].

Rhizospheric microorganism	Compound/extracts	Target strains	Reference
Fungi			
Aspergillus awamori F12	Emodin	*S. aureus*	[75]
		B. subtilis	
Penicillium simplicissimum MA-332	Penicisimpins A–C	*E. coli*	[76]
		Micrococcus luteus	
		P. aeruginosa	
Aspergillus niger MTCC 12676	Ethanol and ethyl acetate extracts	*Streptococcus mutans* (MTCC497)	[77]
		S. aureus (MTCC7443)	
		E. coli (MTCC40)	
		C. albicans (MTCC227)	
		Candida glabrata (MTCC3814)	
Bacteria			
Bacillus pumilus	Bacteriocin-like inhibitory substance (BLIS)	*Listeria monocytogenes* (PTCC 1163)	[78]
		B. cereus (PTCC 1015)	
		S. aureus MRSA (ATCC 1912)	
		Enterococcus VRE	

Rhizospheric microorganism	Compound/extracts	Target strains	Reference
Streptomyces sp. SRDP-H03	Ethyl acetate extract	*S. aureus* (NCIM-2079)	[79]
		B. cereus (NCIM-2016)	
		B. subtilis (NCIM-2699)	
		E. coli (NCIM-2685)	
		K. pneumoniae (NCIM-2957)	
		Vibrio cholerae (MTCC-3905)	
Exiguobacterium mexicanum MSSRFS9	3,6,18-trione, 9,10-dihydro-12 -hydroxyl-2methyl-5-(phenyl methyl)(5-alpha, 10- alpha)-dihydroergotamine (C3) and dipropyl—S-propyl ester (C4)	*E. coli* (ATCC 25922)	[80]
		Shigella flexneri (ATCC 12022)	
		K. pneumonia (ATCC 700603)	
		Salmonella enterica (ATCC 14028)	
Streptomyces sp.	Crude extract	*B. subtilis* (UFPEDA-86)	[81]
	Ethanolic fraction	*S. aureus* (UFPEDA-02)	
	Ethyl acetate fraction	*S. aureus* (MRSA) (UFPEDA-700)	
		C. albicans (UFPEDA-1007)	
Micromonospora sp. A2	- Ethyl acetate extract; - FT-IR included aldehydes, alkynes, 2 aromatic rings, alkanes and alkynes	*S. aureus* MRSA	[82]
Pantoea agglomerans	1-Octadecane and 1-nonadecanol	*Klebsiella* sp. *S. aureus* *S. pneumonia*	[83]
Streptomyces strain M7	Actinomycins	*S. aureus* MRSA (MTCC 96)	[84]
		Enterococcus VRE	
Streptomyces sp. VITBKA3	Ethyl acetate extract	*S. aureus* MRSA (ATCC 43300)	[85]
	(1,1-Dichloropentane (DCP) (76%) - major compound in partial purification)	*S. aureus* MRSA (ATCC700699)	

Table 2.
Secondary metabolites produced by rhizosphere-derived microorganisms and antimicrobial activity against pathogenic microbes.

family and colonize the host plant's root system and form nodules, increasing biological nitrogen fixation, growth and yield of crops [65, 66]. AMF also plays a crucial role in plant health, increasing the efficiency of mineral uptake to promote growth and suppress pathogens [67, 68]. *Aspergillus*, *Fusarium*, *Penicillium*, *Verticillium*, and *Trichoderma* are among the most common fungi genera in the soil [69, 70].

Due to its fundamental function in suppressing pathogens, as well as endophytes, rhizospheric fungi and bacteria, these microorganisms have attracted the attention of researchers as a new source of valuable bioactive metabolites with antimicrobial activity [71–73]. Since antibiotic resistance is a serious global health concern [74], exploring the potential of these microorganisms to discover novel medicine is also of great urgency. In this way, in recent years, secondary metabolites partially or totally identified from microorganisms that inhabit the rhizosphere have been shown to possess antimicrobial activities against important pathogen agents. **Table 2** provides an overview of selected studies that represent significant advances in the search for secondary metabolites produced from rhizospheric fungi and bacteria tested against resistant and multidrug-resistant microorganisms.

Therefore, these and other studies emphasize the vital importance of continuing scientific research to find new antimicrobials and other compounds produced from rhizosphere microorganisms for other biotechnological purposes.

4. Actinobacteria and natural antimicrobial products

Actinobacteria phyla have a high G + C DNA content and share both the characteristics of bacteria and fungi. These Gram-positive filamentous bacteria belong to one of the largest taxonomic groups recognized in the Bacteria domain, widely distributed across ecosystems [86–88].

In terms of metabolite production, the *Streptomyces* genus (**Figure 3**) stands out from other microorganisms due to its variety of bioactive substances and secondary metabolites of economic interest, since more than 80% of the industrially produced antibiotics are processed by this group of microorganisms [89–91].

Streptomyces tubercidicus is known to produce tubercidin, a potent substance that can inhibit several metabolic processes, including pathogens, such as *Trypanosoma cruzi*, viruses, fungi, and present a cytotoxic activity. However, few studies have been done on the isolation of *S. tubercidicus* and only four have been published in the production of bioactive substances [92, 93]. Ratti [94] endophytically isolated the strain of *Streptomyces tubercidicus* (RND-C) from *Solanum lycocarpum* Saint Hill, a medicinal plant typically found in the Brazilian tropical savannah, known for its anti-inflammatory properties. The fractions of the Natural product extract showed high antibiotic activity against *E. coli* and *S. aureus*.

The development of biofilm inhibitors has become a priority in recent years. Bacterial biofilms can tolerate antibiotics and host defense systems, leading to the emergence of drug-resistant and totally drug-resistant infections. As previously mentioned, *Acinetobacter baumannii* leads the list of priority pathogens resistant to antibiotics; therefore, biofilm inhibitors can be applied to decrease antibiotic tolerance by bacteria [95–97]. In this context, [96] conducted a study involving a mutasynthetic approach. Wild-type of *Streptomyces gandocaensis*, isolated from the marine sediment of the island of Punta Mona, in Costa Rica, was ribosome-engineered based on a streptomycin-resistant phenotypes of *S. gandocaensis*, resulting in the activation and improvement of the production of active metabolites. The results showed a production of new substances called cahuitamycins, a peptidic metabolite that showed a potent inhibition in the formation of the biofilm produced by *Acinetobacter baumannii*.

Other studies report different strategies to successfully induce secondary metabolism and, subsequently, produce compounds that are not produced under usual growing conditions. Cryptic genes consist of silent sequences of DNA that are not expressed during the life cycle of a microorganism and can occur through mutations and recombination processes in a few members of a population [98–100].

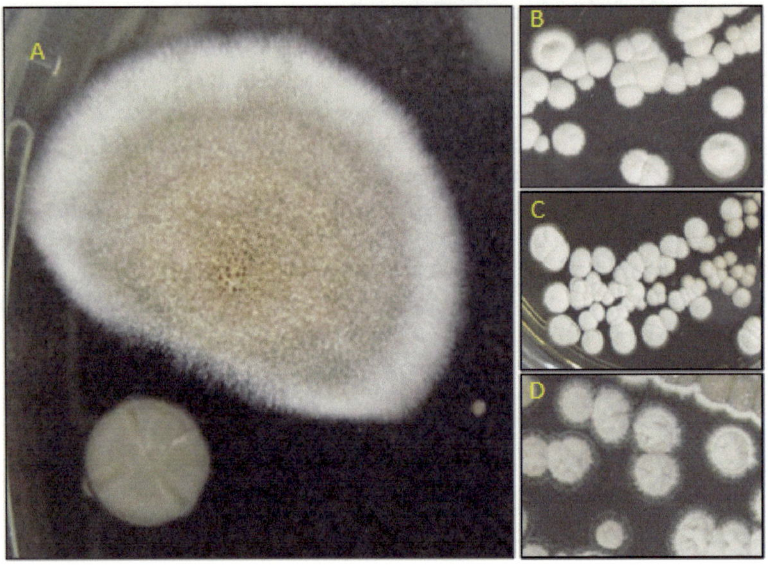

Figure 3.
(A) *Antifungal activity produced by the endophytic Streptomyces sp. during the isolation.* (B–D) *Diversity of rhizospheric streptomycete colonies.*

In this context, cultured actinobacteria combined with mycolic acid-containing bacteria (*Rhodococcus erythropolis*, *Dietzia* spp., *Nocardia* spp., *Williamsia* spp., *Gordonia* spp., *Mycobacterium* spp., and *Corynebacterium* spp.) has been a useful approach for the discovery of antimicrobial natural products [99, 101–103]. However, [102] suggests that mycolic acid is insufficient to activate these cryptic genes in *Streptomyces lividans* under monoculture conditions. According to the report, the direct attachment of *S. lividans* cells on the mycolic acid-containing bacteria is crucial for the successful activation of secondary metabolism.

Caraballo-Rodríguez [3] tested the endophytic actinobacteria *Streptomyces cattleya* RLe1, *S. mobaraensis* RLe3, *S. albospinus* RLe7, *Streptomyces* sp. RLe9 and *Kytasatospora cystarginea* RLe10 co-cultured with endophytic fungi *Coniochaeta* sp. FLe4 and *Colletotrichum boninense* isolated from the Brazilian medicinal plant *Lychnophora ericoides*. The authors identified the broad-spectrum angucycline derived from *S. mobaraensis* and two molecules produced by endophytic fungi.

As already mentioned, the process of antibiotic resistance is spreading rapidly in relation to the discovery of new compounds and their introduction into clinical practice. The CDC classifies pathogens such as *B. anthracis* as biohazard category A, whose infection is fatal, and the symptoms may be similar to a common cold [104]. The preliminary study by [105] involved the isolation of the endophytic and rhizospheric microbiome associated with the medicinal plant *Polygala* sp. Natural products extracts produced by rhizoplane-derived actinomycetes showed potent inhibition against *A. baumannii*, *B. anthracis*, *E. coli* CFT073, *L. monocytogenes*, MR *S. aureus*, *S. enterica*, and *S. flexneri*.

Caryocar brasiliense, known as Pequi, is a tree native to the Brazilian savannah and commonly used in folk medicine. Bioactive substances such as gallic acid, quinic acid, ellagic acid, glucogalin, and corilagin were found in its extracts. In addition, they show a growth inhibition rate of the phytopathogenic *Alternaria solani* [106]. A rhizospheric strain of *Streptomyces* sp. was isolated from *C. brasiliense*, whose crude extract obtained from the axenic cultivation was able to inhibit *C. albicans*; in contrast, the co-cultured *Streptomyces* sp. extract increased the growth of *C. albicans* in 50% and promoted the inhibition of *S. aureus* [107].

Biotechnologically, the *Streptomyces* genus is known to be a skilled producer of a wide range of bioactive substances and represents an unexplored reservoir of unique chemical structures.

5. Natural products and endophytic fungi

The scientific interest in fungal natural products gained notoriety after the paclitaxel discovery [108]. Endophytic fungi exhibit the ability to synthesize plant-derived compounds by mimicking the metabolic pathways of the host plant, which confers multifaceted applications in the fields of agriculture, medicine, and pharmaceuticals [109].

The medicinal plant barbatimão (*Stryphnodendron adstringens*) has healing properties, antimicrobial, antioxidant, and anti-inflammatory activities, and its bark has a rich tannin-content [107, 110]. The study by [111] investigated the antimicrobial and anticancer activities of several fungi isolated from *S. adstringens*. The extract of *Nigrospora oryzae* promoted antifungal activity and inhibited the growth of *C. albicans* and *C. sphaerospermum*, while the extracts of *Diaporthe phaseolorum* and *Xylaria* spp. presented anticancer activities.

Although toxic to humans and animals, mycotoxins are secondary metabolites known for their cytotoxic effect against malignant cells [112]. Several species of *Fusarium* and *Beauveria bassiana* are skilled producers of mycotoxins, such as Beauvericin, which promote apoptosis in mammalian cells and exhibit insecticidal properties [113, 114], while Ochratoxin A is produced by some species of fungi, such as *Aspergillus* spp. and *Penicillium* spp. [115, 116].

The superbug methicillin-resistant *Staphylococcus aureus* is responsible for higher mortality rates in the community and hospital-acquired infections [117] due to its ability to resist multiple classes of antibiotics [118, 119]. In this context, fungal alkaloids are known for their potent antibacterial, anticancer, antiparasitic, and insecticidal activities [120]. In [121], a novel alkaloid compound, GKK1032C, is reported, which is produced by *Penicillium* sp. endophytically associated with the mangrove plant, exhibiting potent activity against methicillin-resistant *S. aureus*.

Saponins exhibit a wide range of biological activities, such as antifungal, hemolytic, antiviral, and immunomodulatory. These compounds represent an alternative to overcome multidrug-resistant microorganisms since they can act synergistically with antibiotics. Moreover, medicines that were once considered ineffective due to resistance problems might be effective for resistant microbes [122, 123]. Nevertheless, as reported by [124], saponin from *Quillaja saponaria* bark did not present synergistic activity in combination with ampicillin, streptomycin, and ciprofloxacin against a clinical strain of *E. coli*. In a short communication from [125], the isolation of triterpenoid saponins produced by the endophytic fungi *Fusarium oxysporum* and *Aspergillus niger* isolated from *Panax notoginseng* was reported. According to the authors, saponin extracts exhibited moderate to high antimicrobial activity against the pathogens tested.

6. Concluding remarks

Antibiotic-resistant microbes represent a severe threat to the public health system worldwide. Furthermore, multidrug-resistant 'ESKAPE' organisms (*Enterococcus* spp., *Staphylococcus aureus*, *Klebsiella* spp., *Acinetobacter baumannii*, *Pseudomonas aeruginosa* and *Enterobacter* spp) are strictly associated with high rates of morbidity and mortality, as well as an economic impact.

In this chapter, we highlighted the strategies of antimicrobial drug discovery produced by endophytes and rhizospheric microorganisms, since enormous untapped resources remain. The use of such microbes in biotechnological processes has increased in recent years, as they are skilled producers of natural bioactive products that can be used as pharmaceuticals to face this ever-increasing threat.

Conflict of interest

The authors declare no conflict of interest.

Funding

This manuscript was financially supported by Sao Paulo State Research Support Foundation (FAPESP) through project number 2016/13423-5.

Author details

Felipe de Paula Nogueira Cruz[1,2], Andréa Cristina Bogas[1,2] and Cristina Paiva de Sousa[1,2]*

1 Laboratory of Microbiology and Biomolecules—LaMiB, Department of Morphology and Pathology, Federal University of São Carlos, Brazil

2 Biotechnology Graduate Program, Federal University of Sao Carlos, Brazil

*Address all correspondence to: prokarya@ufscar.br

References

[1] Lewis K, Epstein S, D'Onofrio A, Ling LL. Uncultured microorganisms as a source of secondary metabolites. Journal of Antibiotics (Tokyo). 2010;**63**(8):468-476. DOI: 10.1038/ja.2010.87

[2] Piza ACMT, Hokka C, Sousa C. Endophytic actinomycetes from *Miconia albicans* (Sw.) Triana (Melastomataceae) and evaluation of its antimicrobial activity. Journal of Science Research and Reports. 2015;**4**(4):281-291. DOI: 10.9734/JSRR/2015/13237

[3] Caraballo-Rodríguez AM, Dorrestein PC, Pupo MT. Molecular inter-kingdom interactions of endophytes isolated from *Lychnophora ericoides*. Scientific Reports. 2017;**7**(1):5373. Published: July 14, 2017. DOI: 10.1038/s41598-017-05532-5

[4] Nicolaou KC, Rigol S. A brief history of antibiotics and select advances in their synthesis. Journal of Antibiotics (Tokyo). 2018;**71**(2):153-184. DOI: 10.1038/ja.2017.62

[5] Chen CH, Kuo HY, Hsu PJ, et al. Clonal spread of carbapenem-resistant *Acinetobacter baumannii* across a community hospital and its affiliated long-term care facilities: A cross sectional study. Journal of Microbiology, Immunology, and Infection. 2018;**51**(3):377-384. DOI: 10.1016/j.jmii.2017.08.001

[6] Fischbach MA, Walsh CT. Antibiotics for emerging pathogens. Science. 2009;**325**(5944):1089-1093. DOI: 10.1126/science.1176667

[7] Walsh CT, Wencewicz TA. Prospects for new antibiotics: A molecule-centered perspective. Journal of Antibiotics (Tokyo). 2014;**67**(1):7-22. DOI: 10.1038/ja.2013.49

[8] Nair DR, Chen J, Monteiro JM, et al. A quinolinol-based small molecule with anti-MRSA activity that targets bacterial membrane and promotes fermentative metabolism. Journal of Antibiotics (Tokyo). 2017;**70**(10):1009-1019. DOI: 10.1038/ja.2017.79

[9] Huggins WM, Minrovic BM, Corey BW, et al. 1,2,4-Triazolidine-3-thiones as narrow spectrum antibiotics against multidrug-resistant *Acinetobacter baumannii*. ACS Medicinal Chemistry Letters. 2016;**8**(1):27-31. Published: November 12, 2016. DOI: 10.1021/acsmedchemlett.6b00296

[10] Sommer MOA, Munck C, Toft-Kehler RV, Andersson DI. Prediction of antibiotic resistance: Time for a new preclinical paradigm? Nature Reviews. Microbiology. 2017;**15**(11):689-696. DOI: 10.1038/nrmicro.2017.75

[11] Joseph B, Pryia MR. Bioactive compounds from endophytes and their potential in pharmaceutical effect: A review. American Journal of Biochemistry and Molecular Biology. 2011;**1**(3):291-309. DOI: 10.3923/ajbmb.2011.291.309

[12] Matsumoto A, Takahashi Y. Endophytic actinomycetes: Promising source of novel bioactive compounds. Journal of Antibiotics (Tokyo). 2017;**70**(5):514-519. DOI: 10.1038/ja.2017.20

[13] Vigliotta G, Giordano D, Verdino A, et al. New compounds for a good old class: Synthesis of two B-lactam bearing cephalosporins and their evaluation with a multidisciplinary approach. Bioorganic & Medicinal Chemistry. 2020;**28**(4):115302. DOI: 10.1016/j.bmc.2019.115302

[14] Azevedo JL, Maccheroni W Jr, Pereira JO, De Araújo WL. Endophytic microorganisms: A review on insect control and recent advances on tropical plants. Electronic Journal of

Biotechnology. 2000;**3**:15-16. DOI: 10.2225/vol3-issue1fulltext-4

[15] Pacifico D, Squartini A, Crucitti D, et al. The role of the Endophytic microbiome in the grapevine response to environmental triggers. Frontiers in Plant Science. 2019;**10**:1256. Published: October 9, 2019. DOI: 10.3389/fpls.2019.01256

[16] Philippot L, Raaijmakers JM, Lemanceau P, van der Putten WH. Going back to the roots: The microbial ecology of the rhizosphere. Nature Reviews. Microbiology. 2013;**11**(11):789-799. DOI: 10.1038/nrmicro3109

[17] Ling LL, Schneider T, Peoples AJ, et al. A new antibiotic kills pathogens without detectable resistance [published correction appears in Nature. 2015 Apr 16;520(7547):388]. Nature. 2015;**517**(7535):455-459. DOI: 10.1038/nature14098

[18] Tong WY, Darah I, Latiffah Z. Antimicrobial activities of endophytic fungal isolates from medicinal herb *Orthosiphon stamineus* Benth. Journal of Medicinal Plant Research: Planta Medica. 2011;**5**:831-836

[19] Serrano NFG. Purificação e caracterização bioquímica de substâncias bioativas produzidas por endofítico isolado de *Prunus* spp. Dissertation. Sao Carlos, Sao Paulo, Brazil: Federal University of São Carlos. 2009

[20] Gouda S, Das G, Sen SK, Shin HS, Patra JK. Endophytes: A treasure house of bioactive compounds of medicinal importance. Frontiers in Microbiology. 2016;**7**:1538. Published: September 2016, 29. DOI: 10.3389/fmicb.2016.01538

[21] Kandel SL, Joubert PM, Doty SL. Bacterial endophyte colonization and distribution within plants. Microorganisms. 2017;**5**(4):77. Published: November 25, 2017. DOI: 10.3390/microorganisms5040077

[22] White JF, Kingsley KL, Zhang Q, et al. Review: Endophytic microbes and their potential applications in crop management. Pest Management Science. 2019;**75**(10):2558-2565. DOI: 10.1002/ps.5527

[23] Alvin A, Miller KI, Neilan BA. Exploring the potential of endophytes from medicinal plants as sources of antimycobacterial compounds. Microbiological Research. 2014;**169** (7-8):483-495. DOI: 10.1016/j.micres.2013.12.009

[24] Stierle A, Strobel G, Stierle D. Taxol and taxane production by *Taxomyces andreanae*, an endophytic fungus of Pacific yew. Science. 1993;**260**(5105):214-216. DOI: 10.1126/science.8097061

[25] Strobel G, Daisy B, Castillo U, Harper J. Natural products from endophytic microorganisms. Journal of Natural Products. 2004;**67**(2):257-268. DOI: 10.1021/np030397v

[26] Pan F, Su TJ, Cai SM, Wu W. Fungal endophyte-derived *Fritillaria unibracteata* var. *wabuensis*: diversity, antioxidant capacities in vitro and relations to phenolic, flavonoid or saponin compounds. Scientific Reports. 2017;**7**:42008. Published: February 6, 2017. DOI: 10.1038/srep42008

[27] Vaz ABM, Brandão LR, Vieira MLA, Pimenta RS, Morais PB, Sobral MEG, et al. Diversity and antimicrobial activity of fungal endophyte communities associated with plants of Brazilian savanna ecosystems. African Journal of Microbiology Research. 2012;**6**(13):3173-3185. DOI: 10.5897/AJMR11.1359

[28] Marcellano JP, Collanto AS, Fuentes RG. Antibacterial activity of endophytic fungi isolated from the bark of *Cinnamomum mercadoi*. The Pharmacogenomics Journal.

2017;**9**(3):405-409. DOI: 10.5530/pj.2017.3.69

[29] Dang L, Li G, Yang Z, et al. Chemical constituents from the endophytic fungus Trichoderma ovalisporum isolated from *Panax notoginseng*. Annales de Microbiologie. 2010;**60**:317-320 https://doi.org/10.1007/s13213-010-0043-2

[30] Wang QX, Li SF, Zhao F, et al. Chemical constituents from endophytic fungus *Fusarium oxysporum*. Fitoterapia. 2011;**82**(5):777-781. DOI: 10.1016/j.fitote.2011.04.002

[31] Sebastianes FL, Cabedo N, El Aouad N, et al. 3-hydroxypropionic acid as an antibacterial agent from endophytic fungi *Diaporthe phaseolorum*. Current Microbiology. 2012;**65**(5):622-632. DOI: 10.1007/s00284-012-0206-4

[32] Subban K, Subramani R, Johnpaul M. A novel antibacterial and antifungal phenolic compound from the endophytic fungus *Pestalotiopsis mangiferae*. Natural Product Research. 2013;**27**(16):1445-1449. DOI: 10.1080/14786419.2012.722091

[33] Ratnaweera PB, Williams DE, de Silva ED, Wijesundera RL, Dalisay DS, Andersen RJ. Helvolic acid, an antibacterial nortriterpenoid from a fungal endophyte, *Xylaria* sp. of orchid *Anoectochilus setaceus* endemic to Sri Lanka. Mycology. 2014;**5**(1):23-28. DOI: 10.1080/21501203.2014.892905

[34] Ibrahim SRM, Elkhayat ES, Mohamed GA, Khedr AIM, Fouad MA, Kotb MHR, et al. Aspernolides F and G, new butyrolactones from the endophytic fungus *Aspergillus terreus*. Phytochemistry Letters. 2015;**14**:84-90 http://doi.org/10.1016/j.phytol.2015.09.006

[35] Ratnaweera PB, de Silva ED, Wijesundera RLC, Andersen RJ. Antimicrobial constituents of *Hypocrea virens*, an endophyte of the mangrove-associate plant *Premna serratifolia* L. Journal of the National Science Foundation of Sri Lanka. 2016;**44**:43-51. DOI: 10.4038/jnsfsr.v44i1.7980

[36] He Y, Hu Z, Sun W, et al. Spiroaspertrione a, a bridged spirocyclic meroterpenoid, as a potent potentiator of oxacillin against methicillin-resistant *Staphylococcus aureus* from *Aspergillus* sp. TJ23. The Journal of Organic Chemistry. 2017;**82**(6):3125-3131. DOI: 10.1021/acs.joc.7b00056

[37] Qiao Y, Zhang X, He Y, et al. Aspermerodione, a novel fungal metabolite with an unusual 2,6-dioxabicyclo[2.2.1]heptane skeleton, as an inhibitor of penicillin-binding protein 2a. Scientific Reports. 2018;**8**(1):5454. Published: April 3, 2018. DOI: 10.1038/s41598-018-23817-1

[38] Hu S, Liang M, Mi Q, et al. Two new diphenyl ether derivatives from the fermentation products of the endophytic fungus *Phomopsis asparagi*. Chemistry of Natural Compounds. 2019;**55**:843-846. DOI: 10.1007/s10600-019-02828-y

[39] Astuti P, Rollando R, Wahyuono S, Nurrochmad A. Antimicrobial activities of isoprene compounds produced by an endophytic fungus isolated from the leaves of *Coleus amboinicus* Lour. Journal of Pharmacy and Pharmacognosy Research. 2020;**8**(4):280-289

[40] Ding L, Maier A, Fiebig HH, Lin WH, Hertweck C. A family of multicyclic indolosesquiterpenes from a bacterial endophyte. Organic & Biomolecular Chemistry. 2011;**9**(11):4029-4031. DOI: 10.1039/c1ob05283g

[41] Ding L, Maier A, Fiebig HH, Lin WH, Peschel G, Hertweck C, et al. Eudesmenes from an endophytic Streptomyces sp. of the mangrove tree

Kandelia candel. Journal of Natural Products. 2012;**75**(12):2223-2227. DOI: 10.1021/np300387n

[42] Djinni I, Defant A, Kecha M, Mancini I. Antibacterial polyketides from the marine alga-derived endophitic *Streptomyces sundarbansensis*: A study on hydroxypyrone tautomerism. Marine Drugs. 2013;**11**(1):124-135. Published: January 10, 2013. DOI: 10.3390/md11010124

[43] Yang X, Peng T, Yang Y, et al. Antimicrobial and antioxidant activities of a new benzamide from endophytic *Streptomyces* sp. YIM 67086. Natural Product Research. 2015;**29**(4):331-335. DOI: 10.1080/14786419.2014.945174

[44] Khieu TN, Liu MJ, Nimaichand S, et al. Characterization and evaluation of antimicrobial and cytotoxic effects of *Streptomyces* sp. HUST012 isolated from medicinal plant *Dracaena cochinchinensis* Lour. Frontiers in Microbiology. 2015;**6**:574. Published 2015 Jun 8. DOI: 10.3389/fmicb.2015.00574

[45] Alshaibani MM, Jalil J, Sidik NM, Edrada-Ebel R, Zin NM. Isolation and characterization of cyclo-(tryptophanyl-prolyl) and chloramphenicol from *Streptomyces* sp. SUK 25 with antimethicillin-resistant *Staphylococcus aureus* activity. Drug Design, Development and Therapy. 2016;**10**:1817-1827. Published: May 31, 2016. DOI: 10.2147/DDDT.S101212

[46] Gos FMWR, Savi DC, Shaaban KA, et al. Antibacterial activity of endophytic actinomycetes isolated from the medicinal plant *Vochysia divergens* (Pantanal, Brazil). Frontiers in Microbiology. 2017;**8**:1642. Published: September 6, 2017. DOI: 10.3389/fmicb.2017.01642

[47] Vu HT, Nguyen DT, Nguyen HQ, et al. Antimicrobial and cytotoxic properties of bioactive metabolites produced by *Streptomyces cavourensis* YBQ59 isolated from *Cinnamomum cassia* Prels in Yen Bai Province of Vietnam. Current Microbiology. 2018;**75**(10):1247-1255. DOI: 10.1007/s00284-018-1517-x

[48] Bunbamrung N, Intaraudom C, Dramae A, et al. Antibacterial, antitubercular, antimalarial and cytotoxic substances from the endophytic *Streptomyces* sp. TBRC7642. Phytochemistry. 2020;**172**:112275. DOI: 10.1016/j.phytochem.2020.112275

[49] Hartmann A, Rothballer M, Schmid M. Lorenz Hiltner, a pioneer in rhizosphere microbial ecology and soil bacteriology research. Plant and Soil. 2008;**312**:7-14. DOI: 10.1007/s11104-007-9514-z

[50] Parray JA, Mir MY, Shameen N. Rhizosphere engineering and agricultural productivity. In: Sustainable Agriculture: Biotechniques in Plant Biology. 2019. DOI: 10.1007/978-981-13-8840-8

[51] Sabale SN, Suryawanshi PP, Krishnaraj PU. Soil Metagenomics: Concepts and Applications, Metagenomics. In: Hozzein WN, editor. Basics, Methods and Applications. IntechOpen; 2019. DOI: 10.5772/intechopen.88958

[52] Brahmaprakash GP, Sahu PK, Lavanya G, Nair SS, Gangaraddi VK, Gupta A. Microbial Functions of the Rhizosphere. In: Singh D, Singh H, Prabha R, editors. Plant-Microbe Interactions in Agro-Ecological Perspectives; Singapore: Springer; 2019. DOI: 10.1007/978-981-10-5813-4_10

[53] Dini-Andreote F, Gumiere T, Durrer A. Exploring interactions of plant microbiomes. Science in Agriculture. 2014;**71**:528-539. DOI: 10.1590/0103-9016-2014-0195

[54] Jones D, Nguyen C, Finlay DR. Carbon flow in the rhizosphere: Carbon trading at the soil–root interface. Plant and Soil. 2009;**321**:5-33. DOI: 10.1007/s11104-009-9925-0

[55] Hassan MK, McInroy JA, Kloepper JW. The interactions of rhizodeposits with plant growth-promoting rhizobacteria in the rhizosphere: A review. Agriculture. 2019;**9**(7):142. DOI: 10.3390/agriculture9070142

[56] Venturi V, Keel C. Signaling in the rhizosphere. Trends in Plant Science. 2016;**21**(3):187-198. DOI: 10.1016/j.tplants.2016.01.005

[57] Kuan KB, Othman R, Abdul Rahim K, Shamsuddin ZH. Plant growth-promoting rhizobacteria inoculation to enhance vegetative growth, nitrogen fixation and nitrogen remobilisation of maize under greenhouse conditions. PLoS One. 2016;**11**(3):e0152478. DOI: 10.1371/journal.pone.0152478

[58] Mehta P, Walia A, Kulshrestha S, Chauhan A, Shirkot CK. Efficiency of plant growth-promoting P-solubilizing *Bacillus circulans* CB7 for enhancement of tomato growth under net house conditions. Journal of Basic Microbiology. 2015;**55**(1):33-44. DOI: 10.1002/jobm.201300562

[59] Gupta G, Parihar SS, Ahirwar NK, Snehi SK, Singh V. Plant growth promoting rhizobacteria (PGPR): Current and future prospects for development of sustainable agriculture. Journal of Microbial and Biochemical Technology. 2015;**7**:096-102. DOI: 10.4172/1948-5948.1000188

[60] Patel T, Saraf M. Biosynthesis of phytohormones from novel rhizobacterial isolates and their in vitro plant growth-promoting efficacy. Journal of Plant Interactions. 2017;**12**:480-487. DOI: 10.1080/17429145.2017.1392625

[61] Berendsen RL, Pieterse CM, Bakker PA. The rhizosphere microbiome and plant health. Trends in Plant Science. 2012;**17**(8):478-486. DOI: 10.1016/j.tplants.2012.04.001

[62] Yadav S, Singh K, Chandra R. Plant Growth–Promoting Rhizobacteria (PGPR) and Bioremediation of Industrial Waste. In: Chandra R, Sobti RC, editor. Microbes for Sustainable Development and Bioremediation. Boca Raton: CRC Press; 2019. DOI: 10.1201/9780429275876

[63] Kour D, Rana KL, Yadav N, Yadav AN, Kumar A, Meena VS, et al. Rhizosphere microbiomes: Biodiversity, mechanisms of plant growth promotion, and biotechnological applications for sustainable agriculture. In: Kumar A, Meena V, editors. Plant Growth Promoting Rhizobacteria for Agricultural Sustainability. Singapura: Springer; 2019. DOI: 10.1007/978-981-13-7553-8_2

[64] Shastri B, Kumar R. Microbial secondary metabolites and plant microbe communications in the rhizosphere. In: Singh, J.S. New and Future Developments in Microbial Biotechnology and Bioengineering. Microbes in Soil, Crop and Environmental Sustainability. B.V: Elsevier; 2019. pp. 93-111. DOI: 10.1016/B978-0-12-818258-1.00006-6

[65] Alam F, Bhuiyan MA, Alam SS, Waghmode TR, Kim PJ, Lee YB. Effect of rhizobium sp. BARIRGm901 inoculation on nodulation, nitrogen fixation and yield of soybean (*Glycine max*) genotypes in gray terrace soil. Bioscience, Biotechnology, and Biochemistry. 2015;**79**(10):1660-1668. DOI: 10.1080/09168451.2015.1044931

[66] Getahun A, Muleta D, Assefa F, Kiros S. Field application of Rhizobial inoculants in enhancing faba bean production in acidic soils: An innovative strategy to improve crop productivity. In: Akhtar M, editor. Salt Stress, Microbes, and Plant Interactions: Causes and Solution. Singapore: Springer; 2019. DOI: 10.1007/978-981-13-8801-9

[67] Amballa H, Bhumi NR. Significance of arbuscular mycorrhizal fungi and

rhizosphere microflora in plant growth and nutrition. In: Choudhary et al., editors. Plant-Microbe Interaction: An Approach to Sustainable Agriculture. Singapura: Springer; 2016. pp. 417-452. DOI: 10.1007/978-981-10-2854-0

[68] Singh I, Giri B. Arbuscular mycorrhiza mediated control of plant pathogens. In: Varma A, Prasad R, Tuteja N, editors. Mycorrhiza—Nutrient Uptake, Biocontrol, Ecorestoration. Cham: Springer; 2017. DOI: 10.1007/978-3-319-68867-1

[69] Brahmaprakash GP, Sahu PK, Nair GLSS, Gangaraddi VK, Gupta A. Microbial functions of the rhizosphere. In: Singh DP, Singh HB, Prabha R, editors. Plant-Microbe Interactions in Agro-Ecological Perspectives. Springer; 2017. DOI: 10.1007/978-981-10-5813-4

[70] Pattnaik SS, Busi S. Rhizosphere fungi: Diversity and potential biotechnological applications. In: Yadav A, Misha S, Singh S, Gupta A, editors. Recent Advancement in White Biotechnology Through Fungi. Fungal Biology. Cham: Springer; 2019. DOI: 10.1007/978-3-030-10480-1

[71] Iniyan AM, Kannan RR, Vincent SGP. Characterization of culturable actinomycetes associated with halophytic rhizosphere as potential source of antibiotics. Proceedings of the National Academy of Sciences, India Section B: Biological Sciences. 2015;**87**:233-242. DOI: 10.1007/s40011-015-0601-2

[72] Muleta A, Assefa F. Isolation and screening of antibiotic producing actinomycetes from rhizosphere and agricultural soils. African Journal of Biotechnology. 2018;**17**:700-714. DOI: 10.5897/AJB2017.16080

[73] Zhang TY, Wu YY, Zhang MY, Cheng J, Dube B, et al. New antimicrobial compounds produced by *Seltsamia galinsogisoli* sp. nov., isolated from *Galinsoga parvifora* as potential inhibitors of FtsZ. Scientific Reports. 2019;**9**:8319. DOI: 10.1038/s41598-019-44810-2

[74] Ahmad M, Khan AU. Global economic impact of antibiotic resistance: A review. Journal of Global Antimicrobial Resistance. 2019;**19**:313-316. DOI: 10.1016/j.jgar.2019.05.024

[75] Chang M, Wang J, Tian F, Zhang Q, Ye B. Antibacterial activity of secondary metabolites from Aspergillus awamori F12 isolated from rhizospheric soil of Rhizophora stylosa Griff. Chinese: Wei Sheng Wu Xue Bao; Oct 2010;**50**(10):1385-1391. PMID: 21141475

[76] Xu R, Li XM, Wang BG. Penicisimpins A–C, three new dihydroisocoumarins from Penicillium simplicissimum MA-332, a marine fungus derived from the rhizosphere of the mangrove plant *Bruguiera sexangula* var. *rhynchopetala*. Phytochemistry Letters. 2016;**17**:114-118. DOI: 10.1016/j.phytol.2016.07.003

[77] Singh A, Kumar M, Salar RK. Isolation of a novel antimicrobial compounds producing fungus *Aspergillus Niger* MTCC 12676 and evaluation of its antimicrobial activity against selected pathogenic microorganisms. Journal of Pure and Applied Microbiology. 2017;**11**(3):1457-1464. DOI: 10.22207/JPAM.11.3.29

[78] Zaghian S, Shokri D, Emtiazi G. Co-production of a UV-stable bacteriocin-like inhibitory substance (BLIS) and indole-3-acetic acid hormone (IAA) and their optimization by Taguchi design in *Bacillus pumilus*. Annales de Microbiologie. 2011;**62**:1189-1197. DOI: 10.1007/s13213-011-0359-6

[79] Rakesh KN, Junaid S, Dileep N, Kekuda PTR. Antibacterial and antioxidant activities of *Streptomyces* species SRDP-H03 isolated

from soil of Hosudi, Karnataka, India. Journal of Drug Delivery Science and Technology. 2013;(4):47-53. DOI: 10.22270/jddt.v3i4.568

[80] Shanthakumar SP, Duraisamy P, Vishwanath G, Selvanesan BC, Ramaraj V, Vasantharaj David B. Broad spectrum antimicrobial compounds from the bacterium *Exiguobacterium mexicanum* MSSRFS9. Microbiological Research. 2015;**178**:59-65. DOI: 10.1016/j.micres.2015.06.007

[81] Silva-Lacerda GR, Santana RC, Vicalvi-Costa MC, et al. Antimicrobial potential of actinobacteria isolated from the rhizosphere of the Caatinga biome plant *Caesalpinia pyramidalis* Tul. Genetics and Molecular Research. 2016;**15**(1):15017488. Published: March 4, 2016. DOI: 10.4238/gmr.15017488

[82] Abdullahi U, Obidah JS, Jada SM. Characterization of antibiotics inhibitory to methicillin resistant *Staphylococcus aureus* (MRSA) from soil actinomycetes. Asian Journal of Research in Medical and Pharmaceutical Sciences. 2018;**4**(2):1-13. DOI: 10.9734/AJRIMPS/2018/39742

[83] Nair NM, Kanthasamy R, Mahesh R, Selvam SIK, Ramalakshmi S. Production and characterization of antimicrobials from isolate *Pantoea agglomerans* of Medicago sativa plant rhizosphere soil. Journal of Applied and Natural Sciences. 2019;**11**(2):267-272. DOI: 10.31018/ jans.v11i2.203

[84] Sharma M, Manhas RK. Purification and characterization of actinomycins from Streptomyces strain M7 active against methicillin resistant *Staphylococcus aureus* and vancomycin resistant *Enterococcus*. BMC Microbiology. 2019;**19**(1):44. Published: February 19, 2019. DOI: 10.1186/s12866-019-1405-y

[85] Bhakyashree K, Kannabiran K. Actinomycetes mediated targeting of drug resistant MRSA pathogens. Journal of King Saud University—Science. 2020;**32**:260-264. DOI: 10.1016/j.jksus.2018.04.034

[86] Barka EA, Vatsa P, Sanchez L, et al. Taxonomy, physiology, and natural products of actinobacteria [published correction appears in Microbiol Mol Biol Rev. 2016 Nov 9;80(4): iii]. Microbiology and Molecular Biology Reviews. 2015;**80**(1):1-43. Published: November 25, 2015. DOI: 10.1128/MMBR.00019-15

[87] Anandan R, Dharumadurai D, Manogaran GP. An Introduction to Actinobacteria. In: Dhanasekaran D, Jiang Y, editors. Actinobacteria - Basics and Biotechnological Applications. IntechOpen; 2016. DOI: 10.5772/62329

[88] Azman AS, Mawang CI, Khairat JE, AbuBakar S. Actinobacteria-a promising natural source of anti-biofilm agents. International Microbiology. 2019;**22**(4):403-409. DOI: 10.1007/s10123-019-00066-4

[89] Bérdy J. Bioactive Microbial Metabolites [published correction appears in J Antibiot (Tokyo). 2005 Apr;58(4):C-1]. Journal of Antibiotics (Tokyo). 2005;**58**(1):1-26. DOI: 10.1038/ja.2005.1

[90] Qin S, Xing K, Jiang JH, Xu LH, Li WJ. Biodiversity, bioactive natural products and biotechnological potential of plant-associated endophytic actinobacteria. Applied Microbiology and Biotechnology. 2011;**89**(3):457-473. DOI: 10.1007/s00253-010-2923-6

[91] Genilloud O. Actinomycetes: Still a source of novel antibiotics. Natural Product Reports. 2017;**34**(10):1203-1232. DOI: 10.1039/c7np00026j

[92] Smulson ME, Suhadolnik RJ. The biosynthesis of the 7-deazaadenine ribonucleoside, tubercidin, by *Streptomyces tubercidicus*. The

Journal of Biological Chemistry. 1967;**242**(12):2872-2876

[93] Kónya A, Szabó Z, Láng I, Barta I, Salát J. Production of FK520 by *Streptomyces tubercidicus*. Microbiological Research. 2008;**163**(6):624-632. DOI: 10.1016/j.micres.2006.10.002

[94] Ratti RP, Piza ACMT, Malpass AC, Hokka CO, Dubreuil JD, Sousa CP. Growing kinetics and antimicrobial activity of Streptomyces tubercidicus crude extracts. In: Microorganisms in Industry and Environment from Scientific and Industrial Research to Consumer Products. Vol. 1. Singapore: World Scientific Publishing Company Pvt Ltd. (Org.); 2010. pp. 589-592

[95] Böttcher T, Kolodkin-Gal I, Kolter R, Losick R, Clardy J. Synthesis and activity of biomimetic biofilm disruptors. Journal of the American Chemical Society. 2013;**135**(8):2927-2930. DOI: 10.1021/ja3120955

[96] Park SR, Tripathi A, Wu J, et al. Discovery of cahuitamycins as biofilm inhibitors derived from a convergent biosynthetic pathway. Nature Communications. 2016;**7**:10710. Published: February 16, 2016. DOI: 10.1038/ncomms10710

[97] Sharma D, Misba L, Khan AU. Antibiotics versus biofilm: An emerging battleground in microbial communities. Antimicrobial Resistance and Infection Control. 2019;**8**:76. Published: May 16, 2019. DOI: 10.1186/s13756-019-0533-3

[98] Tamburini E, Mastromei G. Do bacterial cryptic genes really exist? Research in Microbiology. 2000;**151**(3):179-182. DOI: 10.1016/s0923-2508(00)00137-6

[99] Onaka H. Novel antibiotic screening methods to awaken silent or cryptic secondary metabolic pathways in actinomycetes. Journal of Antibiotics (Tokyo). 2017;**70**(8):865-870. DOI: 10.1038/ja.2017.51

[100] Chagas FO, Pupo MT. Chemical interaction of endophytic fungi and actinobacteria from *Lychnophora ericoides* in co-cultures. Microbiological Research. 2018;**212-213**:10-16. DOI: 10.1016/j.micres.2018.04.005

[101] Onaka H, Mori Y, Igarashi Y, Furumai T. Mycolic acid-containing bacteria induce natural-product biosynthesis in *Streptomyces* species. Applied and Environmental Microbiology. 2011;**77**(2):400-406. DOI: 10.1128/AEM.01337-10

[102] Asamizu S, Ozaki T, Teramoto K, Satoh K, Onaka H. Killing of mycolic acid-containing bacteria aborted induction of antibiotic production by *Streptomyces* in combined-culture. PLoS One. 2015;**10**(11):e0142372. DOI: 10.1371/journal.pone.0142372

[103] Romano S, Jackson SA, Patry S, Dobson ADW. Extending the "one strain many compounds" (OSMAC) principle to marine microorganisms. Marine Drugs. 2018;**16**(7):244. Published: July 23, 2018. DOI: 10.3390/md16070244

[104] Falcinelli SD, Shi MC, Friedlander AM, Chua J. Green tea and epigallocatechin-3-gallate are bactericidal against *Bacillus anthracis*. FEMS Microbiology Letters. 2017;**364**(12). DOI: 10.1093/femsle/fnx127

[105] Nogueira Cruz FP. Isolation of the Endophytic and Rhizospheric Microbiome Associated with Polygala Spp.: Evaluation of the Biotechnological Potential and Antimicrobial Activity. Thesis, Federal University of Sao Carlos; 2018

[106] Breda CA, Gasperini AM, Garcia VL, et al. Phytochemical analysis and antifungal activity of extracts from leaves and fruit residues of Brazilian

savanna plants aiming its use as safe fungicides. Natural Products and Bioprospecting. 2016;**6**(4):195-204. DOI: 10.1007/s13659-016-0101-y

[107] Assis PC. Bactérias endofíticas isoladas de Caryocar brasiliense: atividade enzimática, antimicrobiana, leishimanicida e co-cultura com microrganismos patogênicos. Dissertation, Federal University of Sao Carlos; 2018

[108] Naik BS. Developments in taxol production through endophytic fungal biotechnology: A review. Oriental Pharmacy and Experimental Medicine. 2019;**19**:1-13. DOI: 10.1007/s13596-018-0352-8

[109] Paramanantham P, Pattnaik S, Siddhardha B. Natural products from endophytic fungi: Synthesis and applications. In: Singh BP, editor. Advances in Endophytic Fungal Research: Present Status and Future Challenges. Cham: Springer International Publishing; 2019. pp. 83-103. DOI: 10.1007/978-3-030-03589-1_5

[110] Torres FL. Isolamento, caracterização e potencial biotecnológico de fungos endofíticos associados à plantas do Cerrado. Dissertation, Federal University of Sao Carlos; 2018

[111] Carvalho CR, Gonçalves VN, Pereira CB, Johann S, Galliza IV, et al. The diversity, antimicrobial and anticancer activity of endophytic fungi associated with the medicinal plant *Stryphnodendron adstringens* (Mart.) Coville (Fabaceae) from the Brazilian savannah. Symbiosis. 2012;**57**:95-107

[112] Loi M, Leonardis S, Mulè G, Logrieco AF, PC. A novel and potentially multifaceted dehydroascorbate reductase increasing the antioxidant systems is induced by beauvericinin tomato. Antioxidants (Basel). 2020;**9**(5):E435. Published: May 16, 2020. DOI: 10.3390/antiox9050435

[113] Taevernier L, Veryser L, Roche N, et al. Human skin permeation of emerging mycotoxins (beauvericin and enniatins). Journal of Exposure Science & Environmental Epidemiology. 2016;**26**(3):277-287. DOI: 10.1038/jes.2015.10

[114] Mallebrera B, Prosperini A, Font G, Ruiz MJ. In vitro mechanisms of beauvericin toxicity: A review. Food and Chemical Toxicology. 2018;**111**:537-545. DOI: 10.1016/j.fct.2017.11.019

[115] Vega FE, Posada F, Peterson SW, Gianfagna TJ, Chaves F. Penicillium species endophytic in coffee plants and ochratoxin a production. Mycologia. 2006;**98**(1):31-42. DOI: 10.3852/mycologia.98.1.31

[116] Mondani L, Palumbo R, Tsitsigiannis D, Perdikis D, Mazzoni E, Battilani P. Pest management and ochratoxin a contamination in grapes: A review. Toxins (Basel). 2020;**12**(5):E303. Published: May 7, 2020. DOI: 10.3390/toxins12050303

[117] Pervaiz A, Khan R, Anwar F, Mushtaq G, Kamal MA, Khan H. Alkaloids: An emerging antibacterial modality against methicillin resistant *Staphylococcus aureus*. Current Pharmaceutical Design. 2016;**22**(28):4420-4429. DOI: 10.2174/1381612822999160629115627

[118] Lakhundi S, Zhang K. Methicillin-resistant *Staphylococcus aureus*: Molecular characterization, evolution, and epidemiology. Clinical Microbiology Reviews. 2018;**31**(4) e00020-18. Published: September 12, 2018. DOI: 10.1128/CMR.00020-18

[119] Turner NA, Sharma-Kuinkel BK, Maskarinec SA, et al. Methicillin-resistant Staphylococcus aureus: An overview of basic and clinical research.

Nature Reviews. Microbiology. 2019;**17**(4):203-218. DOI: 10.1038/s41579-018-0147-4

[120] Newmister SA, Gober CM, Romminger S, et al. OxaD: A versatile indolic nitrone synthase from the marine-derived fungus *Penicillium oxalicum* F30. Journal of the American Chemical Society. 2016;**138**(35):11176-11184. DOI: 10.1021/jacs.6b04915

[121] Qi X, Li X, Zhao J, et al. GKK1032C, a new alkaloid compound from the endophytic fungus *Penicillium* sp. CPCC 400817 with activity against methicillin-resistant *S. aureus*. Journal of Antibiotics (Tokyo). 2019;**72**(4):237-240. DOI: 10.1038/s41429-019-0144-5

[122] Liu J, Yang X, He J, Xia M, Xu L, Yang S. Structure analysis of triterpene saponins in *Polygala tenuifolia* by electrospray ionization ion trap multiple-stage mass spectrometry. Journal of Mass Spectrometry. 2007;**42**(7):861-873. DOI: 10.1002/jms.1210

[123] Tagousop CN, Tamokou JD, Kengne IC, Ngnokam D, Voutquenne-Nazabadioko L. Antimicrobial activities of saponins from Melanthera elliptica and their synergistic effects with antibiotics against pathogenic phenotypes. Chemistry Central Journal. 2018;**12**(1):97. Published: September 20, 2018. DOI: 10.1186/s13065-018-0466-6

[124] Arabski M, Węgierek-Ciuk A, Czerwonka G, Lankoff A, Kaca W. Effects of saponins against clinical *E. coli* strains and eukaryotic cell line. Journal of Biomedicine & Biotechnology. 2012;**2012**:286216. DOI: 10.1155/2012/286216

[125] Jin Z, Gao L, Zhang L, et al. Antimicrobial activity of saponins produced by two novel endophytic fungi from *Panax notoginseng*. Natural Product Research. 2017;**31**(22):2700-2703. DOI: 10.1080/14786419.2017.1292265

Chapter 5

Antimicrobial Effect of Titanium Dioxide Nanoparticles

Carol López de Dicastillo, Matias Guerrero Correa, Fernanda B. Martínez, Camilo Streitt and Maria José Galotto

Abstract

The widespread use of antibiotics has led to the emergence of multidrug-resistant bacterial strains, and therefore a current concern for food safety and human health. The interest for new antimicrobial substances has been focused toward metal oxide nanoparticles. Specifically, titanium dioxide (TiO_2) has been considered as an attractive antimicrobial compound due to its photocatalytic nature and because it is a chemically stable, non-toxic, inexpensive, and Generally Recognized as Safe (GRAS) substance. Several studies have revealed this metal oxide demonstrates excellent antifungal and antibacterial properties against a broad range of both Gram-positive and Gram-negative bacteria. These properties were significantly improved by titanium dioxide nanoparticles (TiO_2 NPs) synthesis. In this chapter, latest developments on routes of synthesis of TiO_2 NPs and antimicrobial activity of these nanostructures are presented. Furthermore, TiO_2 NPs favor the inactivation of microorganisms due to their strong oxidizing power by free radical generation, such as hydroxyl and superoxide anion radicals, showing reductions growth against several microorganisms, such as *Escherichia coli* and *Staphylococcus aureus*. Understanding the main mechanisms of antimicrobial action of these nanoparticles was the second main purpose of this chapter.

Keywords: titanium dioxide, nanoparticles, green synthesis, antimicrobial activity

1. Introduction

The incidence of microbial attack in different sectors such as food, textiles, medicine, water disinfection, and food packaging leads to a constant trend in the search for new antimicrobial substances. The increased resistance of some bacteria to some antibiotics and the toxicity to the human body of some organic antimicrobial substances has increased the interest in the development of inorganic antimicrobial substances. Among these compounds, metal and metal oxide compounds have attracted significant attention due to their broad-spectrum antibacterial activities. On the other hand, nanoscale materials are well known thanks to their increased properties due to their high surface area-to-volume ratio. Antimicrobial NPs have shown excellent and different activities from their bulk properties [1, 2].

During last decades, metal oxide nanoparticles, such as zinc oxide (ZnO), manganese oxide (MgO), titanium dioxide (TiO_2), and iron oxide (Fe_2O_3), have been extensively applicable thanks to their unique physiochemical properties in biological applications. Among metal oxide antimicrobial agents, TiO_2 is a valuable semiconducting transition metal oxide material and shows special features, such as easy control, reduced cost, non-toxicity, and good resistance to chemical erosion, that allow its application in optics, solar cells, chemical sensors, electronics, antibacterial and antifungal agents [3]. In general, TiO_2 nanoparticles (TiO_2 NPs) present large surface area, excellent surface morphology, and non-toxicity in nature. Several authors have reported that TiO_2 NPs have been one of the most studied NPs thanks to their photocatalytic antimicrobial activity, exerting excellent bio-related activity against bacterial contamination [4–7].

Antimicrobial activity of nanoparticles is highly influenced by several intrinsic factors such as their morphology, size, chemistry, source, and nanostructure [8–11]. Specifically, antimicrobial activity of TiO_2 NPs is greatly dependent on photocatalytic performance of TiO_2, which depends strongly on its morphological, structural, and textural properties [12]. Several TiO_2 NPs have been developed through different methods of synthesis. Specifically, in this chapter, eco-friendly synthesis based on biological sources, such as natural plant extracts and metabolites from microorganisms, which have resulted in TiO_2 NPs with different size, shape, morphology, and crystalline structures will be presented. Titanium dioxide produces amorphous and crystalline forms and primarily can occur in three crystalline polymorphous: anatase, rutile, and brookite. Studies on synthesis have stated that the crystalline structure and morphology of TiO_2 NPs is influenced by process parameters such as hydrothermal temperatures, starting concentration of acids, etc. [13]. The crystal structures and the shape of TiO_2 NPs are both the most important properties that affect their physicochemical properties, and therefore their antimicrobial properties [14]. Regarding the crystal structures, anatase presents the highest photocatalytic and antimicrobial activity. Some works have shown that anatase structure can produce $OH^{\cdot}$ radicals in a photocatalytic reaction, and as it will be clearly explained below, bacteria wall and membranes can be deadly affected [15, 16].

2. Antimicrobial activity of titanium dioxide NPs

2.1 Latest tendencies on TiO_2 nanoparticle synthesis

The potential health impact and toxicity to the environment of NPs is currently an important matter to be addressed. Several works have confirmed that metal oxide NPs conventionally synthesized using chemical methods, such as sol–gel synthesis and chemical vapor deposition, have shown different levels of toxicity to test organisms [17–20]. In recent years, researchers have emphasized on the development of nanoparticles promoted through environmental sustainability and processes characterized by an ecological view, mild reaction conditions, and non-toxic precursors. Due to this growing sensitivity toward green chemistry and biological processes, ecological processes are currently being investigated for the synthesis of non-toxic nanoparticles.

These biological methods are considered safe, cost-effective, biocompatible, non-toxic, sustainable, and environmentally friendly processes [20]. Furthermore, it has been described that chemically synthesized NPs have exhibited less stability and added agglomeration, resulting in biologically synthesized NPs that are more dispersible, stable in size, and the processes consuming less energy [21].

These biosynthetic methods, also called "green synthesis," use various biological resources available in nature, including live plant [22], plant products, plant extracts, algae, fungi, yeasts [23], bacteria [24], and virus for the synthesis of NPs. Among these methods, the processes that use plant-based materials are considered the most suitable for large-scale green synthesis of NPs with respect to their ease and safety [25]. On the other hand, the reduction rate of metal ions in the presence of the plant extract is much faster compared to microorganisms, and provides stable particles [26]. Plants contain biomolecules that have been highly studied by researchers like phenols, nitrogen compounds, terpenoids, and other metabolites. It is well known that the hydroxyl and carboxylic groups present in these biocompounds act as stabilizers and reducing agents due to their high antioxidant activity [12]. Thus, plant extracts have been studied as one of the best green alternatives for metal oxide nanoparticles synthesis [27]. In recent years, TiO_2 nanoparticles have been obtained by using different plant extracts, but not all of them have been studied for their antimicrobial activity. **Table 1** presents a compilation of synthesized TiO2 nanoparticles from green synthesis by using plant extracts that were tested against different microorganisms.

Different factors need to be evaluated in this research field in order to obtain TiO_2 NPs with better properties and to maintain their biocompatibility. It has been shown that nanoparticles obtained from green synthesis can have a better morphology and size translated into better antimicrobial activity. Mobeen and Sundaram have obtained TiO_2 NPs from titanium tetrachloride precursor through a chemical and a green synthesis method. Sulfuric acid and ammonium hydroxide were used

Source	Titanium precursor	Size (nm)	Shape/ crystal structure	Target microorganism (method)
Azadirachta indica leaves extract [28]	TiO_2	25–87 (SEM)	Spherical/ anatase-rutile	*S. typhi*, *E. coli*, and *K. pneumoniae* (broth micro dilution method)
Psidium guajava leaves extract [29]	$TiO(OH)_2$	32.58 (FESEM)	Spherical shape and clusters/ anatase-rutile	*S. aureus* and *E. coli* (agar diffusion)
Vitex negundo Linn leaves extract [30]	$Ti\{OCH(CH_3)_2\}_4$	26–15 (TEM)	Spherical and rod shaped/ tetragonal phase anatase	*S. aureus* and *E. coli* (agar diffusion)
Morinda citrifolia leaves extract [31]	$TiCl_4$	15–19 (SEM)	Quasi-spherical shape/rutile	*S. aureus*, *B. subtilis*, *E. coli*, *P. aeruginosa*, *C. albicans*, *A. niger* (agar diffusion)
Trigonella foenum-graecum leaf extract [21]	$TiOSO_4$	20–90 (HR-SEM)	Spherical/ anatase	*E. faecalis*, *S. aureus*, *S. faecalis*, *B. subtilis.*, *Y. enterocolitica*, *P. vulgaris*, *E. coli*, *P. aeruginosa*, *K. pneumoniae*, and *C. albicans* (agar diffusion)
Orange peel extract [32]	$TiCl_4$	20–50 (SEM)	Irregular and angular structure with high porous net/ anatase	*S. aureus*, *E. coli*, and *P. aeruginosa* (agar diffusion)
Glycyrrhiza glabra root extracts [33]	TiO_2	60–140 (FESEM)	Spherical shape/ anatase	*S. aureus* and *K. pneumoniae* (agar diffusion)

Table 1.
Synthesis of TiO_2 NPs by using plant extracts.

in the chemical-based method and, in the green synthesis, those chemical reagents were replaced by an orange peel extract [32]. The nanoparticles obtained by using the natural extract presented a well-defined and smaller crystalline nature (approx. 17.30 nm) compared to the nanoparticles synthesized through the chemical method (21.61 nm). Both methods resulted in anatase crystalline structures, and, when evaluating the antimicrobial activity, the more eco-friendly NPs revealed higher bactericidal activity against Gram-positive and Gram-negative bacteria compared to the chemically synthesized nanoparticles.

Bavanilatha et al. have also detailed TiO_2 NPs green synthesis with *Glycyrrhiza glabra* root extract. Antibacterial activity against *Staphylococcus aureus* and *Klebsiella pneumonia* were investigated and *in vivo* toxicity tests using the zebrafish embryonic model (*Danio rerio*) were also carried out [33]. Results have demonstrated their biocompatibility because healthy embryos of adult fish to different variations of NP and no distinctive malformations were observed at every embryonic stage with respect to embryonic controls.

Subhapriya and Gomathipriya have biosynthesized TiO_2 NPs by using *a Trigonella foenum-graecum* leaf extract, obtaining spherical NPs and their size varied between 20 and 90 nm, and their antimicrobial activity was evaluated through the standard method of disc diffusion [21]. The NPs showed significant antimicrobial activity against *Yersinia enterocolitica* (10.6 mm), *Escherichia coli* (10.8 mm), *Staphylococcus aureus* (11.2 mm), *Enterococcus faecalis* (11.4 mm), and *Streptococcus faecalis* (11.6 mm). Results confirmed developed TiO_2 NPs as an effective antimicrobial drug that can lead to the progression of new antimicrobial drugs.

Spherical TiO_2 NPs were synthesized from plants, in particular by applying a *Morinda citrifolia* leaf extract, and through advanced hydrothermal method [31]. Developed TiO_2 NPs showed a size between 15 and 19 nm in an excellent quasispherical shape. In addition, their antimicrobial activity was tested against human pathogens, such as *Staphylococcus aureus*, *Escherichia coli, Bacillus subtilis, Pseudomonas aeruginosa, Candida albicans,* and *Aspergillus niger*. TiO_2 NPs exhibited interesting antimicrobial activity, principally against Gram-positive bacteria.

In addition to plants, other organisms can produce inorganic compounds at an intra or extracellular level. The synthesis of TiO_2 NPs through microorganisms, including bacteria, fungi, and yeasts, also meets the requirements and the exponentially growing technological demand toward eco-friendly strategies, by avoiding the use of toxic chemicals in the synthesis and protocols [34]. The metabolites generated by microorganism present bioreducing, capping, and stabilizing properties that improve the NPs synthesis performance. Jayaseelan et al. have stated glycyl-L-proline, one of the most abundant metabolite from *Aeromonas hydrophilia* bacteria, as the main compound that acted as a capping and stabilizing agent during TiO_2 NPs green synthesis [35]. Moreover, the interest in fungi in green synthesis of metal oxide nanoparticles has increased over last years. Fungi enzymes and/or metabolites also present intrinsically the potential to obtain elemental or ionic state metals from their corresponding salts [34, 36]. Different works based on the green synthesis of TiO_2 NPs from bacteria and fungus are presented in **Table 2**. Some of them have been synthesized with antimicrobial and antifungal purposes, and their target microorganisms are also declared.

Two important factors that affect NPs synthesis are the type of microorganisms and their source. Some microorganisms widely used in the food industry are *Lactobacillus*, a bacterium used in dairy products and as a probiotic supplement, and *Saccharomyces cerevisiae,* a yeast commonly used in bakery. Jha et al. have investigated the effectiveness of both microorganisms to synthesize TiO_2 NPs. A comparison between synthesis through *Lactobacillus* from yogurt and probiotic tablets resulted in different NP sizes: a particle size of 15–70 nm for yogurt, and 10–25 nm

Microorganism	Titanium precursor	Size (nm)	Shape/crystal structure	Target microorganisms (method)
Aeromonas hydrophilia [46]	$TiO(OH)_2$	28–54 (SEM) ~ 40.5 (XRD)	Spherical/uneven	*S. aureus, S. pyogenes* (agar diffusion)
Aspergillus flavus [34]	TiO_2	62–74 (TEM)	Spherical/anatase and rutile	*E. coli, P. aeruginosa, K. pneumoniae, B. subtilis* (agar diffusion and MIC)
Bacillus mycoides [37]	Titanyl hydroxide	40–60 (TEM)	Spherical/anatase	*E. coli* (toxicity)
Bacillus subtilis [38]	K_2TiF_6	11–32 (TEM)	Spherical	Aquatic biofilm
Fusarium oxysporum [36]	K_2TiF_6	6–13 (TEM)	Spherical/brookite	—
Lactobacillus sp. [51]	$TiO(OH)_2$	~ 24.6 (TEM)	Spherical/ anatase-rutile	—
Planomicrobium sp. [39]	TiO_2	100–500 (SEM)	Irregular/pure crystalline	*B. subtilis, K. planticola, Aspergillus niger* (agar diffusion)
Propionibacterium jensenii [52]	$TiO(OH)_2$, 300°C	15–80 (FESEM)	Spherical	—
Saccharomyces cerevisiae [51]	$TiO(OH)_2$	~ 12.6 (TEM)	Spherical/ anatase-rutile	—

Table 2.
Examples of TiO_2 NPs synthesis through microorganisms, both bacteria and fungus strains.

for tablets. This difference was due to the purity of the bacteria [40]. In general, TiO_2 NP synthesis through microorganisms has not provided stable sizes, being not industrially scalable compared to the synthesis of nanoparticles from plants.

2.2 Antimicrobial activity of TiO_2 NPs

Harmful bacteria, such as *Staphylococcus aureus, Burkholderia cepacia, Pseudomonas aeruginosa, Clostridium difficile, Klebsiella pneumoniae, Escherichia coli, Acinetobacter baumannii, Mycobacterium tuberculosis,* and *Neisseria gonorrhoeae,* are responsible for bacterial infections that can cause serious diseases in humans year after year [40]. The principal solution is the use of antibiotics, antimicrobial and antifungal agents. Nevertheless, in recent years there has been an increase in the resistance of several bacterial strains to these substances, and therefore there is currently a great interest in the search for new antimicrobial substances. The antimicrobial nanoparticles have been studied due to their high activity, specifically the metal oxide nanoparticles [41–43]. In this sense, titanium dioxide nanoparticles are one of the antimicrobial NPs whose study has gained interest during last years.

TiO_2 is a thermally stable and biocompatible chemical compound with high photocatalytic activity and has presented good results against bacterial contamination [44]. **Table 3** presents some research including the antimicrobial capacity of TiO_2 NPs.

Microorganism	NPs	Results
Methicillin-resistant *Staphylococcus aureus* [45]	Fe_3O_4-TiO_2 core/shell magnetic NPs	The survival ratio [%] of bacteria decreased from 82.40 to 7.13%.
Staphylococcus saprophyticus [45]	Fe_3O_4-TiO_2 core/shell magnetic NPs	The survival ratio [%] of bacteria decreased from 79.15 to 0.51%.
Streptococcus pyogenes [57]	Fe_3O_4-TiO_2 core/shell magnetic NPs	The survival ratio [%] of bacteria decreased from 82.87 to 4.45%.
Escherichia coli [46]	TiO_2 nanotubes ~ 20 nm	97.53% of reduction
Staphylococcus aureus [46]	TiO_2 nanotubes ~ 20 nm	99.94% of reduction
Bacillus subtilis [47]	TiO_2 NPs co-doped with silver (19–39 nm)	1% Ag-N-TiO2 had the highest antibacterial activity with antibacterial diameter reduction of 22.8 mm
Mycobacterium smegmatis [48]	Cu-doped TiO_2NPs ~20 nm	The percentage of inhibition was around 47%
Pseudomonas aeruginosa [49]	TiO_2 NPs 10–25 nm	Although it was not completely euthanized, their survival was significantly inhibited.
Shewanella oneidensis MR-1 [48]	Cu-doped TiO_2 NPs ~20 nm	The percentage of inhibition was around 11%

Table 3.
TiO_2 nanoparticles against different microorganisms and their antimicrobial activities.

The principal factors differentiating the antimicrobial activity between TiO_2 NPs were their morphology, crystal nature, and size. According to López de Dicastillo et al. [11], hollow TiO_2 nanotubes presented interesting antimicrobial reduction thanks to the enhancement of specific surface area. This fact can be explained by the nature of titanium dioxide, and one of the main mechanisms of its action is through the generation of reactive oxygen species (ROS) on its surface during the process of photocatalysis when it exposed to light at an appropriate wavelength. It is important to highlight that some research works have evidenced antimicrobial activity of TiO_2 NPs increased when they were irradiated with UV-A light due to the photocatalytic nature of this oxide. The time of irradiation varied between 20 min [45] and 3 hours [50].

3. Understanding the antimicrobial mechanism of TiO_2 NPs toward bacteria

Titanium dioxide nanoparticles (TiO_2 NPs) are one of the most studied materials in the area of antimicrobial applications due to its particular abilities, such as bactericidal photocatalytic activity, safety, and self-cleaning properties. The mechanism referred to the antimicrobial action of TiO_2 is commonly associated to reactive oxygen species (ROS) with high oxidative potentials produced under band-gap irradiation photo-induces charge in the presence of O_2 [51]. ROS affect bacterial cells by different mechanisms leading to their death. Antimicrobial substances with broad spectrum activity against microorganisms (Gram-negative and Gram-positive bacteria and fungi) are of particular importance to overcome the MDR (multidrug resistance) generated by traditional antibiotic site-specific.

The main photocatalytic characteristic of TiO_2 is a wide band gap of 3.2 eV, which can trigger the generation of high-energy electron–hole pair under UV-A light with wavelength of 385 nm or lower [52]. As mentioned above for bulk powder, TiO_2 NPs have the same mechanism based on the ROS generation with the advantage of being at nanoscale. This nanoscale nature implies an important increase of surface area-to-volume ratio that provides maximum contact with environment water and oxygen [53] and a minimal size, which can easily penetrate the cell wall and cell membrane, enabling the increase of the intracellular oxidative damage.

Bacteria have enzymatic antioxidant defense systems like catalases and superoxide dismutase, in addition to natural antioxidants like ascorbic acid, carotene, and tocopherol, which inhibit lipid peroxidation or O-singlet and the effects of ROS radicals such as $OH_2^{\bullet-}$ and $OH^{\bullet}$. When those systems are exceeded, a set of redox reactions can lead to the death cell by the alteration of different essential structures (cell wall, cell membrane, DNA, etc.) and metabolism routes [54]. In the following sections, several ways that cellular structures were affected in the presence of TiO_2 NPs will be described. In order to understand the genome responses of bacteria to TiO_2 photocatalysis, some biological approaches related to expression of genes encoding to defense and repair mechanism of microorganism will explained below. Different mechanisms and processes of antimicrobial activity of TiO_2 NPs are represented as a global scheme in **Figure 1**.

3.1 Cell wall

ROS are responsible for the damage by oxidation of many organic structures of microorganisms. One of them is the cell wall, which is the first defense barrier against any injury from the environment, thus being the first affected by oxidative damage. Depending on the type of microorganism, the cell wall will have different composition; that is, in fungi and yeast, cell walls are mainly composed of chitin and polysaccharides [55], Gram-positive bacteria contain many layers of peptidoglycan and teichoic acid, and Gram-negative bacteria present a thin layer of peptidoglycan surrounded by a secondary lipid membrane reinforced with transmembrane lipopolysaccharides and lipoproteins [56]. Thus, the effect of TiO_2 NPs will be slightly different depending type of microorganism.

It has been studied that the composition of the cell wall in *Pichia pastoris* (yeast) changed in the presence of TiO_2, increasing the chitin content in response to the

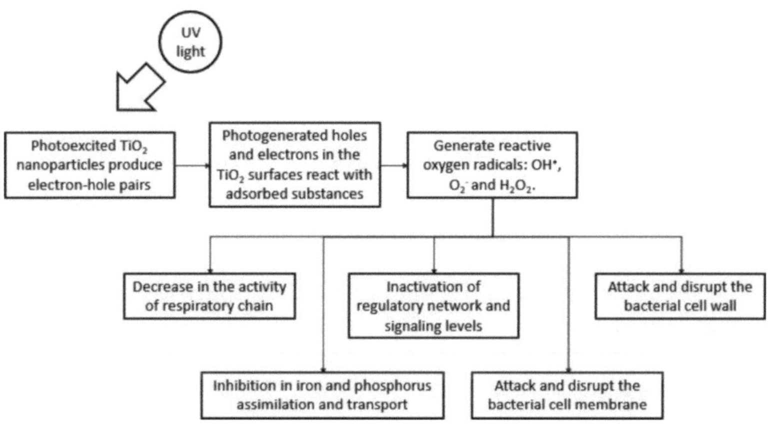

Figure 1.
Scheme of main antimicrobial activity-based processes.

ROS effects [57]. The cell wall of *Escherichia coli* (Gram-negative) composed of lipo-polysaccharide, phosphatidyl-ethanolamine, and peptidoglycan has been reported to be sensitive to the peroxidation caused by TiO_2 [58]. The damage can be quantified by assessing the production of malondialdehyde (MDA), which is a biomarker of lipid peroxidation, or through ATR-FTIR of the supernatant of cell culture, which evidenced the way that porins and proteins on the outer membrane were affected, probably as a result of greater exposure to the surface of TiO_2 [59]. In fungi, the release of $OH^{\bullet}$ captured hydrogen atoms from sugar subunits of polysaccharides, which composed the cell wall, leading to the cleavage of polysaccharide chain and the exposition of cell membrane [60].

In terms of genetic issues, there is evidence that the bacteria change the level expression of certain genes encoding for proteins involved in lipopolysaccharide and peptidoglycan metabolism, pilus biosynthesis, and protein insertion related to the cell wall which values were lower-expressed after exposition to TiO_2 NPs [61].

3.2 Cell membrane

The second usual cellular target of most of antibiotics is the cell membrane mainly composed by phospholipids, which grant the cell a non-rigid cover, permeability, and protection. Most of the studies with TiO_2 NPs have been focused to the loss of membrane integrity caused by oxidation of phospholipids due to ROS such hydroxyl radicals and hydrogen peroxide [62, 63], which led to an increase in the membrane fluidity, leakage of cellular content, and eventually cell lysis.

Gram-positive bacteria present only one membrane protected by many layers of peptidoglycan, whereas Gram-negative bacteria are composed by two membranes, inner and outer, and a thin layer of peptidoglycan between them. The outer membrane is exposed, thus, more liable to mechanical breakage due to the lack of peptidoglycan protective cover, like in Gram-positive bacteria [64]. Some studies have demonstrated a better antimicrobial performance of TiO_2 NPs against Gram-positive bacteria [65] while others reported that Gram-negative bacteria were more resistant [66, 67]. It can be concluded that the bacterial inactivation effectiveness depends mainly on the resistant capacity of cell wall structures and the damage level of ROS generation [68].

In contrast with the lower expression of genes related to the cell wall seen before, the level expression of genes encoding for enzymes involved in metabolism of lipid essential for the cell membrane structure, are over-expressed [61]. It would be concluded that cells compensate the initial cell wall damage by reinforcing the second defense barrier, the cell membrane, in a way to provide support against the oxidation produced by ROS.

In fungi, the biocidal effect is not quite different. In the presence of TiO_2 NPs and UV light, hydroxyl radicals, hydrogen peroxide, and superoxide anions initially promote oxidation of the membrane, leading to an unbalance in the cell permeability, even decomposition of cell walls [69]. This oxidation can inhibit cell respiration by affecting intracellular membranes in mitochondria. Studies have demonstrated biocidal effects on *Penicillium expansum* [70], but there is still research on other strains.

Beyond the relatively well-studied initial lipoperoxidation attack of TiO_2 NPs on the outer/inner cell membrane of the microorganism, specific mechanisms are still aimed of being solved.

3.3 Inhibition of respiratory chain

As the oxidative damage generates lipoperoxidation of cell membranes due to their lipid nature, the respiratory chain, which takes place in the double-membrane mitochondria, is also affected. This organelle is a natural source of ROS in aerobic metabolism because superoxide anions are produced in the electron transfer respiratory chain process. Mitochondria can control this fact by converting them into H_2O_2 by superoxide dismutase (SOD), and finally into water by glutathione peroxidase and catalase [71]. The presence of TiO_2 NPs increases the production of ROS at levels that this enzymatic defense mechanism cannot attenuate the damage, even a dysregulation in electron transfer through the mitochondrial respiratory chain implies an increase in ROS generation [72].

The genetic approaches have indicated that changes in level expression in genes related to the energy production in mitochondria prioritize the most efficient pathway to uptake oxygen, which is through ubiquinol coenzyme [61]. This coenzyme presented a higher capacity to exchange electrons, while the coenzyme-independent oxygen uptake pathways were expressed at lower level.

3.4 DNA

Damage at molecular level in DNA affects all regulatory microorganism metabolism, replication, transcription, and cell division. DNA is particularly sensitive to oxidative damage because oxygen radicals, specially $OH^{\cdot}$ produced by Fenton reaction [73], may attack the sugar-phosphate or the nucleobases and cause saccharide fragmentation aimed to the strand break [74].

DNA strand modifications are more lethal than base modifications (punctual mutation). Mitochondrial DNA is more vulnerable to oxidative damage than nuclear DNA because it is closer to a major cellular ROS source [75].

Besides the enzymatic detoxification system (SOD, glutathione and catalase), DNA injuries are covered by a set of structures related to post-translational modification, protein turnover, chaperones (related to folding), DNA replication and repair, which are significantly over-expressed in the presence of TiO_2 NPs [61].

3.5 Assimilation and transport of iron and inorganic phosphate (Pi)

Iron is an essential ion for cell growth and survival, but it can turn potentially toxic if some malfunction in homeostatic regulation occurs (i.e., Fenton reaction that produces ROS). Bacteria are able to regulate iron concentration in order to maintain it in a physiological range [76]. This regulation involves directly siderophores to active transport of iron in cell [77], whose coding genes related to siderophore synthesis and iron transport protein are significantly lower-expressed in the presence of TiO_2 NPs, decreasing the ability to assimilate and transport it, leading to cell death [61]. The loss of homeostasis regulation was confirmed by ICP-MS analysis, which revealed that the presence of TiO_2 NPs significantly reduced the cellular iron level in *Pseudomonas brassicacearum*, directly proportional to the cell viability [78].

Regarding the functions related to Pi group (PO_4^{3-}) uptake, major differences were found in the expression of set of genes contained in Pho regulon, which were significantly lower when compared to the control [61]. The Pho regulon is a regulatory network in bacteria, yeast, plants, and animals, related to assimilation of inorganic phosphate, merely available in nature, and essential to nutritional cross-talk, secondary metabolite production, and pathogenesis [79].

This suggested that the microorganisms were highly deficient in phosphorus uptake and metabolism in the presence of TiO_2 NPs. It should be also noted that the Pho regulon has been reported to regulate biofilm synthesis capacity and pathogenicity [80].

3.6 Cell-to-cell communication

TiO_2 NPs can directly oxidize components of cell signaling pathways and even change the gene expression by interfering with transcription factors [81]. There is evidence to confirm the interference of TiO_2 NPs in biosynthesis pathways of signaling molecules that bind lipopolysaccharide, stabilize and protect the cell wall against oxidative damage [82]. Moreover, a significant decrease in the synthesis of quorum-sensing signal molecule related to functions like pathogenesis and biofilm development was observed. This was corroborated through Scanning Electron Microscopy (SEM) images of bacteria (*P. aeruginosa*) growth in the presence of TiO_2 NPs without UV irradiation. Cells appeared mainly non-aggregated and dispersed in the substratum, compared with controls without NPs where cells were mainly aggregated by lateral contact. This suggested that TiO_2 NPs not only affected microorganisms by oxidative damage, but also bacteria aggregation and biofilm formation, which directly influenced in pathogenicity [83].

In plants and algae, ROS can act as signaling intermediates in the process of transcription factor controlling stress response by H_2O_2, which is activated by a GSH peroxidase, and not by peroxides directly. But there is still lack of research in this area [84].

4. Conclusions

The control of morphology and crystal structure of TiO_2 NPs is the most important factor to enhance their antimicrobial activity. The appropriate design based on desirable surface properties given by shaped nanoparticles can improve effectiveness that is also dependent on the type of bacteria. The route of synthesis of TiO_2 NPs is also a key factor. Recent works have revealed more eco-friendly synthesis methods, principally based on plant-based compounds and microorganisms, such as bacteria and fungus. Antimicrobial activity of different TiO_2 NPs against Gram-positive and Gram-negative bacteria including antibiotic-resistant strains has been confirmed in different works.

Specific studies on antimicrobial mechanisms have evidenced that microorganism exposed to photocatalytic TiO_2 NPs exhibited cell inactivation at regulatory network and signaling levels, an important decrease in the activity of respiratory chain, and inhibition in the ability to assimilate and transport iron and phosphorous. These processes with the extensive cell wall and membrane alterations were the main factors that explain the biocidal activity of TiO_2 NPs.

Acknowledgements

The authors acknowledge the financial support of CONICYT through the Project Fondecyt Regular 1170624 and "Programa de Financiamiento Basal para Centros Científicos y Tecnológicos de Excelencia" Project FB0807, and CORFO Project 17CONTEC-8367.

Conflict of interest

The authors declare no conflict of interest.

Author details

Carol López de Dicastillo*, Matias Guerrero Correa, Fernanda B. Martínez, Camilo Streitt and Maria José Galotto
Center of Innovation in Packaging (LABEN), CEDENNA (Center for the Development of Nanoscience and Nanotechnology), University of Santiago de Chile (USACH), Santiago, Chile

*Address all correspondence to: analopez.dediscastillo@usach.cl

References

[1] Hajipour MJ, Fromm KM, Akbar Ashkarran A, Jimenez de Aberasturi D, de Larramendi IR, Rojo T, et al. Antibacterial properties of nanoparticles. Trends in Biotechnology. 2012;**30**:499-511. DOI: 10.1016/j.tibtech.2012.06.004

[2] Whitesides G. Nanoscience, nanotechnology, and chemistry. Small. 2005;**1**:172-179. DOI: 10.1002/smll.200400130

[3] Khan SUM, Al-Shahry M, Ingler WB. Efficient photochemical water splitting by a chemically modified n-TiO_2. Science. 2002;**297**:2243-2245. DOI: 10.1126/science.1075035.

[4] Chung IM, Park I, Seung-Hyun K, Thiruvengadam M, Rajakumar G. Plant-mediated synthesis of silver nanoparticles: Their characteristic properties and therapeutic applications. Nanoscale Research Letters. 2016;**11**: 1-14. DOI: 10.1186/s11671-016-1257-4

[5] Bui AKT, Bacic A, Pettolino F. Polysaccharide composition of the fruit juice of *Morinda citrifolia* (noni). Phytochemistry. 2006;**67**:1271-1275. DOI: 10.1016/j.phytochem.2006.04.023

[6] Ravikumar P, Kumar SS. Antifungal activity of extracellularly synthesized silver nanoparticles from *Morinda citrifolia* L. International Journal of Technical Research and Applications. 2014;**2**:108-111

[7] Inbathamizh L, Ponnu TM, Mary EJ. In vitro evaluation of antioxidant and anticancer potential of Morinda pubescens synthesized silver nanoparticles. Journal of Pharmacy Research. 2013;**6**:32-38. DOI: 10.1016/j.jopr.2012.11.010

[8] De Oliveira RC, de Foggi CC, Teixeira MM, Da Silva MDP, Assis M, Francisco EM, et al. Mechanism of antibacterial activity via morphology change of α-AgVO3: Theoretical and experimental insights. ACS Applied Materials & Interfaces. 2017;**9**:11472-11481. DOI: 10.1021/acsami.7b00920

[9] Pal S, Tak YK, Song JM. Does the antibacterial activity of silver nanoparticles depend on the shape of the nanoparticle? A study of the gram-negative bacterium *Escherichia coli*. Applied and Environmental Microbiology. 2007;**73**:1712-1720. DOI: 10.1128/AEM.02218-06

[10] Gilbertson LM, Albalghiti EM, Fishman ZS, Perreault F, Corredor C, Posner JD, et al. Shape-dependent surface reactivity and antimicrobial activity of nano-cupric oxide. Environmental Science & Technology. 2016;**50**:3975-3984. DOI: 10.1021/acs.est.5b05734

[11] López de Dicastillo C, Patiño C, Galotto MJ, Palma JL, Alburquenque D, Escrig J. Novel antimicrobial titanium dioxide nanotubes obtained through a combination of atomic layer deposition and electrospinning technologies. Nanomaterials. 2018;**8**:128. DOI: 10.3390/nano8020128

[12] He Z, Cai Q, Fang H, Situ G, Qiu J, Song S, et al. Photocatalytic activity of TiO_2 containing anatase nanoparticles and rutile nanoflower structure consisting of nanorods. Journal of Environmental Sciences. 2013;**25**:2460-2468. DOI: 10.1016/S1001-0742(12)60318-0

[13] Sarkar D, Ghosh CK, Chattopadhyay KK. Morphology control of rutile TiO_2 hierarchical architectures and their excellent field emission properties. CrystEngComm. 2012;**14**:2683-2690. DOI: 10.1039/c2ce06392a

[14] Burda C, Chen X, Narayanan R, El-Sayed MA. Chemistry and properties

of nanocrystals of different shapes. Chemical Reviews. 2005;**105**:1025-1102. DOI: 10.1021/cr030063a

[15] Zhang Q, Yan X, Shao R, Dai H, Li S. Preparation of nano-TiO_2 by liquid hydrolysis and characterization of its antibacterial activity. Journal Wuhan University of Technology, Materials Science Edition. 2014;**29**:407-409. DOI: 10.1007/s11595-014-0930-7

[16] Vimbela GV, Ngo SM, Fraze C, Yang L, Stout DA. Antibacterial properties and toxicity from metallic nanomaterials. International Journal of Nanomedicine. 2017;**12**:3941-3965. DOI: 10.2147/IJN.S134526

[17] Puzyn T, Rasulev B, Gajewicz A, Hu X, Dasari TP, Michalkova A, et al. Using nano-QSAR to predict the cytotoxicity of metal oxide nanoparticles. Nature Nanotechnology. 2011;**6**:175-178. DOI: 10.1038/nnano.2011.10

[18] He X, Fu P, Aker WG, Hwang HM. Toxicity of engineered nanomaterials mediated by nano–bio–eco interactions. Journal of Environmental Science and Health - Part C Environmental Carcinogenesis and Ecotoxicology Reviews. 2018;**36**:21-42. DOI: 10.1080/10590501.2017.1418793

[19] Hwang HM, Ray PC, Yu H, He X. Toxicology of designer/engineered metallic nanoparticles. In book: Sustainable preparation of metal nanoparticles. 2012:190-212. Chapter 8. DOI: 10.1039/9781849735469-00190

[20] Shah M, Fawcett D, Sharma S, Tripathy SK, Poinern GEJ. Green synthesis of metallic nanoparticles via biological entities. Materials. 2015;**8**:7278-7308. DOI: 10.3390/ma8115377

[21] Subhapriya S, Gomathipriya P. Green synthesis of titanium dioxide (TiO_2) nanoparticles by Trigonella foenum-graecum extract and its antimicrobial properties. Microbial Pathogenesis. 2018;**116**:215-220. DOI: 10.1016/j.micpath.2018.01.027

[22] Bali R, Razak N, Lumb A, Harris AT. The synthesis of metallic nanoparticles inside live plants. In: Proc. 2006 Int. Conf. Nanosci. Nanotechnology, ICONN. 2006. pp. 224-227. DOI: 10.1109/ICONN.2006.340592

[23] Moghaddam AB, Moniri M, Azizi S, Rahim RA, Bin AA, Saad WZ, et al. Biosynthesis of ZnO nanoparticles by a new Pichia kudriavzevii yeast strain and evaluation of their antimicrobial and antioxidant activities. Molecules. 2017;**22**(6):872. DOI: 10.3390/molecules22060872

[24] Jha Z, Behar N, Narayan Sharma S, Chandel G, Sharma D, Pandey MP, et al. Nanotechnology: Prospects of agricultural advancement. Nano Vision. 2011;**1**:88-100

[25] Jose Varghese R, Zikalala N, Sakho EHM, Oluwafemi OS. Green synthesis protocol on metal oxide nanoparticles using plant extracts. Colloidal Metal Oxide Nanoparticles. 2020:67-82. Chapter 5. DOI: 10.1016/B978-0-12-813357-6.00006-1

[26] Nasrollahzadeh M, Maham M, Mohammad Sajadi S. Green synthesis of CuO nanoparticles by aqueous extract of Gundelia tournefortii and evaluation of their catalytic activity for the synthesis of N-monosubstituted ureas and reduction of 4-nitrophenol. Journal of Colloid and Interface Science. 2015;**455**:245-253. DOI: 10.1016/j.jcis.2015.05.045

[27] Nadeem M, Tungmunnithum D, Hano C, Abbasi BH, Hashmi SS, Ahmad W, et al. The current trends in the green syntheses of titanium oxide nanoparticles and their

applications. Green Chemistry Letters and Reviews. 2018;**11**:492-502. DOI: 10.1080/17518253.2018.1538430

[28] Thakur BK, Kumar A, Kumar D. Green synthesis of titanium dioxide nanoparticles using Azadirachta indica leaf extract and evaluation of their antibacterial activity. South African Journal of Botany. 2019;**124**:223-227. DOI: 10.1016/j.sajb.2019.05.024

[29] Santhoshkumar T, Rahuman AA, Jayaseelan C, Rajakumar G, Marimuthu S, Kirthi AV, et al. Green synthesis of titanium dioxide nanoparticles using Psidium guajava extract and its antibacterial and antioxidant properties. Asian Pacific Journal of Tropical Medicine. 2014;7:968-976. DOI: 10.1016/S1995-7645(14)60171-1

[30] Ambika S, Sundrarajan M. [EMIM] BF 4 ionic liquid-mediated synthesis of TiO_2 nanoparticles using Vitex negundo Linn extract and its antibacterial activity. Journal of Molecular Liquids. 2016;**221**:986-992. DOI: 10.1016/j.molliq.2016.06.079

[31] Sundrarajan M, Bama K, Bhavani M, Jegatheeswaran S, Ambika S, Sangili A, et al. Obtaining titanium dioxide nanoparticles with spherical shape and antimicrobial properties using *M. citrifolia* leaves extract by hydrothermal method. Journal of Photochemistry and Photobiology B: Biology. 2017;**171**:117-124. DOI: 10.1016/j.jphotobiol.2017.05.003

[32] Mobeen Amanulla A, Sundaram R. Green synthesis of TiO2 nanoparticles using orange peel extract for antibacterial, cytotoxicity and humidity sensor applications. Materials Today Proceedings. 2019;**8**:323-331. DOI: 10.1016/j.matpr.2019.02.118

[33] Bavanilatha M, Yoshitha L, Nivedhitha S, Sahithya S. Bioactive studies of TiO_2 nanoparticles synthesized using Glycyrrhiza glabra. Biocatalysis and Agricultural Biotechnology. 2019;**19**:101131. DOI: 10.1016/j.bcab.2019.101131

[34] Rajakumar G, Rahuman AA, Roopan SM, Khanna VG, Elango G, Kamaraj C, et al. Fungus-mediated biosynthesis and characterization of TiO2 nanoparticles and their activity against pathogenic bacteria. Spectrochimica Acta Part A: Molecular and Biomolecular Spectroscopy. 2012;**91**:23-29. DOI: 10.1016/J.SAA.2012.01.011

[35] Jayaseelan C, Rahuman AA, Roopan SM, Kirthi AV, Venkatesan J, Kim SK, et al. Biological approach to synthesize TiO_2 nanoparticles using Aeromonas hydrophila and its antibacterial activity. Spectrochimica Acta - Part A: Molecular and Biomolecular Spectroscopy. 2013;**107**:82-89. DOI: 10.1016/j.saa.2012.12.083

[36] Bansal V, Rautaray D, Bharde A, Ahire K, Sanyal A, Ahmad A, et al. Fungus-mediated biosynthesis of silica and titania particles. Journal of Materials Chemistry. 2005;**15**:2583. DOI: 10.1039/b503008k

[37] Órdenes-Aenishanslins NA, Saona LA, Durán-Toro VM, Monrás JP, Bravo DM, Pérez-Donoso JM. Use of titanium dioxide nanoparticles biosynthesized by Bacillus mycoides in quantum dot sensitized solar cells. Microbial Cell Factories. 2014;**13**:90. DOI: 10.1186/s12934-014-0090-7

[38] Dhandapani P, Maruthamuthu S, Rajagopal G. Bio-mediated synthesis of TiO_2 nanoparticles and its photocatalytic effect on aquatic biofilm. Journal of Photochemistry and Photobiology B: Biology. 2012;**110**:43-49. DOI: 10.1016/j.jphotobiol.2012.03.003

[39] Sinica DP, Annadurai G. Pelagia research library novel eco-friendly synthesis of titanium oxide nanoparticles by using *Planomicrobium* sp. and its antimicrobial evaluation. Der Pharmacia. 2013;**4**:59-66

[40] Jha AK, Prasad K. Biosynthesis of metal and oxide nanoparticles using lactobacilli from yoghurt and probiotic spore tablets. Biotechnology Journal. 2010;**5**:285-291. DOI: 10.1002/biot.200900221

[41] Raja S, Ramesh V, Thivaharan V. Green biosynthesis of silver nanoparticles using Calliandra haematocephala leaf extract, their antibacterial activity and hydrogen peroxide sensing capability. Arabian Journal of Chemistry. 2017;**10**:253-261. DOI: 10.1016/j.arabjc.2015.06.023

[42] Müller JC, Botelho GGK, Bufalo AC, Boareto AC, Rattmann YD, Martins ES, et al. *Morinda citrifolia* Linn (noni): In vivo and in vitro reproductive toxicology. Journal of Ethnopharmacology. 2009;**121**:229-233. DOI: 10.1016/j.jep.2008.10.019

[43] Pai AR, Kavitha S, Shweta Raj S, Priyanka P, Vrinda A, Vivin TS, et al. Green synthesis and characterizations of silver nanoparticles using fresh leaf extract of *Morinda citrifolia* and its anti-microbial activity studies. International Journal of Pharmacy and Pharmaceutical Sciences. 2015;7:459-461

[44] Ma W, Li J, Liu Y, Ren X, Gu ZG, Xie Z, et al. Preparation and characterization of excellent antibacterial TiO_2/N-halamines nanoparticles. Colloids and Surfaces A: Physicochemical and Engineering Aspects. 2016;**506**:284-290. DOI: 10.1016/j.colsurfa.2016.06.055

[45] Chen W-J, Tsai P-J, Chen Y-C. Functional Fe_3O_4/TiO_2 core/shell magnetic nanoparticles as photokilling agents for pathogenic bacteria. Small. 2008;**4**:485-491. DOI: 10.1002/smll.200701164

[46] Podporska-Carroll J, Panaitescu E, Quilty B, Wang L, Menon L, Pillai SC. Antimicrobial properties of highly efficient photocatalytic TiO2 nanotubes. Applied Catalysis B: Environmental. 2015;**176-177**:70-75. DOI: 10.1016/j.apcatb.2015.03.029

[47] Yuan Y, Ding J, Xu J, Deng J, Guo J. TiO_2 nanoparticles co-doped with silver and nitrogen for antibacterial application. Journal of Nanoscience and Nanotechnology. 2010;**10**:4868-4874. DOI: 10.1166/jnn.2010.2225

[48] Wu B, Huang R, Sahu M, Feng X, Biswas P, Tang YJ. Bacterial responses to Cu-doped TiO2 nanoparticles. Science of the Total Environment. 2010;**408**:1755-1758. DOI: 10.1016/j.scitotenv.2009.11.004

[49] Tsuang Y-H, Sun J-S, Huang Y-C, Lu C-H, Chang WH-S, Wang C-C. Studies of photokilling of bacteria using titanium dioxide nanoparticles. Artificial Organs. 2008;**32**:167-174. DOI: 10.1111/j.1525-1594.2007.00530.x

[50] López de Dicastillo C, Patiño C, Galotto MJ, Vásquez-Martínez Y, Torrent C, Alburquenque D, et al. Novel hollow titanium dioxide nanospheres with antimicrobial activity against resistant bacteria. Beilstein Journal of Nanotechnology. 2019;**10**:1716-1725. DOI: 10.3762/bjnano.10.167

[51] Verdier T, Coutand M, Bertron A, Roques C. Antibacterial activity of TiO_2 photocatalyst alone or in coatings on *E. coli*: The influence of methodological aspects. Coatings. 2014;**4**:670-686. DOI: 10.3390/coatings4030670

[52] Xie J, Hung YC. Methodology to evaluate the antimicrobial effectiveness of UV-activated TiO_2 nanoparticle-embedded cellulose acetate film.

Food Control. 2019;**106**:106690. DOI: 10.1016/j.foodcont.2019.06.016

[53] Kaladhar Reddy A, Kambalyal PB, Shanmugasundaram K, Rajesh V, Donthula S, Patil SR. Comparative evaluation of antimicrobial efficacy of silver, titanium dioxide and zinc oxide nanoparticles against streptococcus mutans. Pesquisa Brasileira Em Odontopediatria e Clinica Integrada. 2018;**18**:1-8. DOI: 10.4034/PBOCI.2018.181.88

[54] Kiwi J, Rtimi S. Mechanisms of the antibacterial effects of TiO 2 -FeO x under solar or visible light: Schottky barriers versus surface plasmon resonance. Coatings. 2018;**8**:391. DOI: 10.3390/coatings8110391

[55] Gow NAR, Latge J-P, Munro CA. The fungal Cell Wall: Structure, biosynthesis, and function. Microbiology Spectrum. 2017;**5**(3):1-25. DOI: 10.1128/microbiolspec.FUNK-0035-2016

[56] Salton MRJ, Kim K-S. Structure. Galveston: University of Texas Medical Branch; 1996

[57] Liu Z, Zhang M, Han X, Xu H, Zhang B, Yu Q, et al. TiO_2 nanoparticles cause cell damage independent of apoptosis and autophagy by impairing the ROS-scavenging system in Pichia pastoris. Chemico-Biological Interactions. 2016;**252**:9-18. DOI: 10.1016/j.cbi.2016.03.029

[58] Pulgarin C, Kiwi J, Nadtochenko V. Mechanism of photocatalytic bacterial inactivation on TiO_2 films involving cell-wall damage and lysis. Applied Catalysis B: Environmental. 2012;**128**:179-183. DOI: 10.1016/j.apcatb.2012.01.036

[59] Carré G, Hamon E, Ennahar S, Estner M, Lett MC, Horvatovich P, et al. TiO2 photocatalysis damages lipids and proteins in *Escherichia coli*. Applied and Environmental Microbiology. 2014;**80**:2573-2581. DOI: 10.1128/AEM.03995-13

[60] Hammel KE, Kapich AN, Jensen KA, Ryan ZC. Reactive oxygen species as agents of wood decay by fungi. Enzyme and Microbial Technology. 2002;**30**:445-453. DOI: 10.1016/S0141-0229(02)00011-X

[61] Kubacka A, Diez MS, Rojo D, Bargiela R, Ciordia S, Zapico I, et al. Understanding the antimicrobial mechanism of TiO 2 -based nanocomposite films in a pathogenic bacterium. Scientific Reports. 2014;**4**:1-9. DOI: 10.1038/srep04134

[62] Pavlović VP, Vujančević JD, Mašković P, Ćirković J, Papan JM, Kosanović D, et al. Structure and enhanced antimicrobial activity of mechanically activated nano TiO2. Journal of the American Ceramic Society. 2019;**102**:7735-7745. DOI: 10.1111/jace.16668

[63] Khezerlou A, Alizadeh-Sani M, Azizi-Lalabadi M, Ehsani A. Nanoparticles and their antimicrobial properties against pathogens including bacteria, fungi, parasites and viruses. Microbial Pathogenesis. 2018;**123**:505-526. DOI: 10.1016/j.micpath.2018.08.008

[64] Jameel ZN, Mahmood OA, Ahmed FL. Studying the effect of synthesized nano-titanium dioxide via two phases on the *Pseudomonas aeruginosa* and portus bacteria as antimicrobial agents. International Journal of Nanoelectronics and Materials. 2019;**12**:329-338

[65] Cheigh C-I, Park M-H, Chung M-S, Shin J-K, Park Y-S. Comparison of intense pulsed light- and ultraviolet (UVC)-induced cell damage in *Listeria monocytogenes* and *Escherichia coli* O157:H7. Food Control. 2012;**25**:654-659. DOI: 10.1016/J.FOODCONT.2011.11.032

[66] Dunlop PSM, Sheeran CP, Byrne JA, McMahon MAS, Boyle MA, McGuigan KG. Inactivation of clinically relevant pathogens by photocatalytic coatings. Journal of Photochemistry and Photobiology A: Chemistry. 2010;**216**:303-310. DOI: 10.1016/J.JPHOTOCHEM.2010.07.004

[67] van Grieken R, Marugán J, Pablos C, Furones L, López A. Comparison between the photocatalytic inactivation of Gram-positive *E. faecalis* and Gram-negative *E. coli* faecal contamination indicator microorganisms. Applied Catalysis B: Environmental. 2010;**100**:212-220. DOI: 10.1016/J.APCATB.2010.07.034

[68] Zhu Z, Cai H, Sun DW. Titanium dioxide (TiO2) photocatalysis technology for nonthermal inactivation of microorganisms in foods. Trends in Food Science and Technology. 2018;**75**:23-35. DOI: 10.1016/j.tifs.2018.02.018

[69] Li J, Yu H, Wu Z, Wang J, He S, Ji J, et al. Room temperature synthesis of crystalline anatase TiO2 on bamboo timber surface and their short-term antifungal capability under natural weather conditions. Colloids and Surfaces A: Physicochemical and Engineering Aspects. 2016;**508**:117-123. DOI: 10.1016/j.colsurfa.2016.08.045

[70] Maneerat C, Hayata Y. Antifungal activity of TiO2 photocatalysis against Penicillium expansum in vitro and in fruit tests. International Journal of Food Microbiology. 2006;**107**:99-103. DOI: 10.1016/j.ijfoodmicro.2005.08.018

[71] Staerck C, Gastebois A, Vandeputte P, Calenda A, Larcher G, Gillmann L, et al. Microbial antioxidant defense enzymes. Microbial Pathogenesis. 2017;**110**:56-65. DOI: 10.1016/J.MICPATH.2017.06.015

[72] Xue C, Li X, Liu G, Liu W. Evaluation of mitochondrial respiratory chain on the generation of reactive oxygen species and cytotoxicity in HaCaT cells induced by Nanosized titanium dioxide under UVA irradiation. International Journal of Toxicology. 2016;**35**:644-653. DOI: 10.1177/1091581816661853

[73] Gogniat G, Dukan S. TiO2 photocatalysis causes DNA damage via Fenton reaction-generated hydroxyl radicals during the recovery period. Applied and Environmental Microbiology. 2007;**73**:7740-7743. DOI: 10.1128/AEM.01079-07

[74] Imlay J, Linn S. Damage and oxygen radical. Science. 1988;**240**:1302-1309

[75] Ševců A, El-Temsah YS, Joner EJ, Černík M. Oxidative stress induced in microorganisms by zero-valent iron nanoparticles. Microbes and Environments. 2011;**26**:271-281. DOI: 10.1264/jsme2.me11126

[76] Andrews SC, Robinson AK, Rodríguez-Quiñones F. Bacterial iron homeostasis. FEMS Microbiology Reviews. 2003;**27**:215-237. DOI: 10.1016/S0168-6445(03)00055-X

[77] Neilands JB. Siderophores of bacteria and fungi. Microbiological Sciences. 1984;**1**:9-14

[78] Liu W, Bertrand M, Chaneac C, Achouak W. TiO_2 nanoparticles alter iron homeostasis in: *Pseudomonas brassicacearum* as revealed by PrrF sRNA modulation. Environmental Science: Nano. 2016;**3**:1473-1482. DOI: 10.1039/c6en00316h

[79] Santos-Beneit F. The pho regulon: A huge regulatory network in bacteria. Frontiers in Microbiology. 2015;**6**:402. DOI: 10.3389/fmicb.2015.00402

[80] Haddad A, Jensen V, Becker T, Häussler S. The pho regulon influences biofilm formation and type three secretion in *Pseudomonas aeruginosa*. Environmental Microbiology

Reports. 2009;**1**:488-494. DOI: 10.1111/j.1758-2229.2009.00049.x

[81] Apel K, Hirt H. Reactive oxygen species: Metabolism, oxidative stress, and signal transduction. Annual Review of Plant Biology. 2004;**55**:373-399. DOI: 10.1146/annurev.arplant.55.031903.141701

[82] Johnson L, Mulcahy H, Kanevets U, Shi Y, Lewenza S. Surface-localized Spermidine protects the *Pseudomonas aeruginosa* outer membrane from antibiotic treatment and oxidative stress. Journal of Bacteriology. 2012;**194**:813-826. DOI: 10.1128/JB.05230-11

[83] Kubacka A, Serrano C, Ferrer M, Lunsdorf H, Bielecki P, Cerrada ML, et al. High-performance dual-action polymer-TiO2 nanocomposite films via melting processing. Nano Letters. 2007;7:2529-2534. DOI: 10.1021/nl0709569

[84] Ledford HK, Niyogi KK. Singlet oxygen and photo-oxidative stress management in plants and algae. Plant, Cell and Environment. 2005;**28**:1037-1045. DOI: 10.1111/j.1365-3040.2005.01374.x

Chapter 6

Dairy Farms Biosecurity to Protect against Infectious Diseases and Antibiotics Overuse

Stelian Baraitareanu and Livia Vidu

Abstract

Biosecurity is a key element in the battle against antibiotic resistance. The goals of biosecurity are focused not only on the reduction or prevention of the introduction of new diseases from outside sources but also on the reduction or prevention of the movement of infectious diseases on the farm. In this regard, the use of antibiotics can be reduced by simple actions such as physically inspecting animals, testing for bovine diseases, vaccination, or quarantine for at least 3 weeks before mixing with the herd of all new additions. All these examples reduce the risk of diseases with germs from outside. This chapter attempts to synthesize the best biosecurity solutions that can be applied in modern dairy farms.

Keywords: antibioresistance, biosecurity, dairy farm, cattle

1. Introduction

In dairy farms, biosecurity, surveillance, resilience/immunity, biocontainment, and control of disease spread within the herd are the pillars that need to be appropriately managed to ensure the healthy herd [1].

Biosecurity is focused to reduce and prevent the introduction of diseases or pests of animals on a farm, and to minimize the spread of diseases or pests within a farm. Biosecurity action plans need to be implemented mainly in large dairy farms where the disease agents can be introduced by various sources such as labor, advisers, replacement cattle, supplies, feedstuffs, and vehicles [2].

Surveillance programs are developed for early detection of emerging pathogens, to establish disease-free status or the prevalence of a specific disease in a herd [3].

Relation resilience immunity is based on the individuals' resistance to diseases that can be modulated by the ability of animals to adapt to adverse conditions (stress factor) and recover from them [4].

Biocontainment and control programs are important backup systems for biosecurity plans that will prevent the emerging disease spreading within the herd or the endemic diseases spreading between animals into the farm [2, 5].

The overall biosecurity of dairy farm uses different levels or shells of actions (national or supranational, regional, and local), linked with the epidemiological profile of the pathogen. For highly contagious infectious agent (e.g., foot-and-mouth disease), the most efficient biosecurity plan is at national or European Union level, while for infectious agents transmitted by close contact between animals (e.g., bovine tuberculosis), the regional biosecurity measures such as movement controls will protect the status of the region [1].

Biosecurity practices on livestock farms have been described and prioritized in various ways [1, 2, 5, 6]. In this chapter, we grouped biosecurity measures in the following categories: dairy farm sanitation, facility biosecurity, animal biosecurity, feed biosecurity, and manure biosecurity.

2. Dairy farm sanitation

2.1 Employees and visitors

Some infectious agents are specific for dairy cattle and others are zoonotic, affecting both bovine and human health. Employees and visitors can contribute to the spread of all these infectious agents on a dairy farm [7]. The transmission of pathogens by humans can be reduced or even stopped by providing on-farm laundry facilities for all protective clothing used on the farm, using only clean overalls during farm visits, providing disposable clean booties for visitors and cleaning of boots with disinfecting solution after scrubbing off any visible dirt at the end of the visit, and washing of hands before and after working with sick or young animals [7–9].

Milking parlor personnel should wear latex gloves while milking to reduce the spreading potential of the contagious mastitis pathogens [9]. Sometimes, these hired personnel can take care of other animals outside the dairy farm and carry pathogens on the farm. Employees should be regularly trained in good practices to prevent the spread of disease (the principles of hygiene and disease security). They need to know that calves are susceptible to diseases carried by adult animals, and daily activities should be organized so that employees work with younger animals before working with older animals. Prevention of the infectious agent's introduction and spreading from outside and inside sources should also be considered in the education of hired personnel in basic hygiene and disinfection [10]. The main actions included in the biosecurity plan for dairy farms should reduce the risk of infectious diseases to be introduced by employees and visitors (**Table 1**).

The access of visitors must be limited and recorded in a logbook; the farm touring must start from younger to older animal groups; barn doors are recommended to be locked and a warning sign must be posted to keep out unauthorized personnel [9].

Also, along the access road of the farm must be displayed signs directing visitors to the administrative area and to the visitor parking, as well as warning signs to limit direct contact of visitors with farm feed and animals [11].

2.2 Equipment biosecurity

Equipment can be contaminated with infectious secretions, excretions, and blood and the movement of equipment between stalls and farms may also transport pathogens [12].

All equipment used on the farm must be regularly cleaned and disinfected [11]. To prevent contamination of equipment, storage containers need to be used for all tools and feeding equipment. Also, all storage containers are regularly cleaned and disinfected. The storage containers must protect equipment from diseases, pests, or weeds [13]. Before use in healthy animals, equipment that has been used on sick animals must be cleaned and disinfected. However, it is better not to use clothing, shoes, and tools dedicated to the compartment of sick animals [14]. Dehorners, ear

Biosecurity measure	Action
Record in the logbook all farm visitors	Place the visitor logbook at the farm entrance
Restrict the access of visitors to the stable	Locking the stable doors
Inform unauthorized persons that they are not allowed to enter the stable	Post-warning signs asking visitors not to pass inside stable and several directing signs to the farm offices
The visitors can access the stable only with clean clothes and boots, which they have not used in other farms	Provide clean boots and overalls for all visitors
The visitors should use a footbath with disinfectant and clean their boots before entering the stable	Place a disinfectant footbath and brushes outside the stable
The dealer or transporter of the newly arrived animals is not allowed to enter in stable or in contact with the farm animals	The access of the cars is made on a route that avoids contact with the farm animals, directly toward the quarantine area located at a distance from the herd
The livestock renderer access in the stable or the contact with cattle is restricted	Store dead animals away from the stable and main roads

Table 1.
Biosecurity measures designed to reduce the risk of the infectious disease's introduction in dairy farms by employees and visitors.

taggers, hoof knives, clippers, and all shared and hired equipment will be cleaned and disinfected between uses [11, 14].

Nursing bottles and buckets must be sanitized before each feeding [14], calves kept indoors must have fresh clean dry bedding, and plastic calf hutches will be cleaned and disinfected after use [11].

The equipment used for manure disposal will not be used for transporting or delivering feed [13].

Disposable clothing and used veterinary equipment must be removed safely [11].

2.3 Vehicle biosecurity

Vehicles are considered fomites mainly for pathogenic robust organisms that can survive a long time in the environment [1]. Mainly external vehicles that collect milk, calves, and carcasses or deliver feedstuffs, pharmaceuticals, and semen can be involved in the transmission of infectious disease because they travel daily from farm to farm [2, 10]. A high biosecurity risk is associated with carcasses (dead stock) collectors because they are usually in contact with diseased animals [15, 16].

To prevent the introduction of infectious agents, vehicles must be kept clean and should not have access to the zones where the animals are housed [10, 11, 17].

External vehicles should not be allowed on the farm [18]. If vehicles are necessary on the farm, then ensure that vehicles and trailers are clean when entering the farm and disinfected before and after use [6, 11, 18, 19]. Cleaning and disinfection will cover both the exterior and the interior of the vehicles, with greater attention to areas where dirt may be hidden (e.g., wheel arches and tires) [11]. Because the transport by dealers may pose additional risks of infectious disease transmission between farms, it is recommended that the animal moving will use only farm-owned vehicles [20], with clean and ample bedding to prevent both injuries and disease [14].

Guidance indicators and warning and restricting access signs to unauthorized vehicles must be placed at the entrance to the farm road and along the road.

The farm must have a designated area for visitors' vehicles that are at the entrance of the farm and away from the animal and animal stalls [6, 10, 14]. Also, service vehicles should not drive over the routes of feed delivery or manure handling [14].

3. Buildings biosecurity

In a dairy farm, the building's design can help prevent the spread of pathogens to sick cows, periparturient cows, and newborns [2]. Buildings should have a well-established destination, in correlation with the categories of animals present on the farm. Dairy farms can secure their premises against domestic and wild animals by installing various types of fences (e.g., electric fence) around the buildings. Disinfectant footbaths should be at the entry of livestock housing. All farms should have isolation building (the quarantine facility) where the health status of the newly purchased cows will be observed before they join the rest of the herd [21]. To prevent direct and indirect contact between residents and new animals, the quarantine facility should be located in the farthest possible place on the dairy farm [10]. The farm must have a biosecurity plan that includes building maintenance activities (e.g., check and maintain fences, replace bird netting, and repair holes in buildings), which will reduce the contact of cattle with wild animals and the feed contamination with birds droppings or badger feces [14, 21].

4. Animal biosecurity

4.1 Live animal management

The introduction of new cattle is one of the most important biosecurity risks for dairy farms [10]. In modern dairy farming, the sale and movement of cattle is an intrinsic part of the business as a consequence of the increased herd replacement rate of adult milking cows, the forced culling, and the need to increase the size of the herd [1]. Therefore, keeping a closed herd is the most effective biosecurity measure but is the least practical [6]. To reduce the risk of diseases spreading between farms, the new animals are purchased only from herds with known health status and known vaccination protocols [9, 10].

The best solution to prevent the introduction of diseases through the acquisition of new animals is the hosting of the newly purchased cows in a quarantine facility with trained personnel to handle isolated animals [10, 21]. Quarantine is one of the most important biosecurity tools and consists of the separation of specific groups of animals to prevent the transmission of infectious diseases. Prophylactic quarantine is designed to separate the resident herd from newly acquired animals for 1 month or more. During the 30 days of isolation, the personnel from the quarantine facility will monitor cattle health status and prevent direct and indirect contact between new and resident animals [9, 10]. If the infections have short incubation times, then the animals will develop acute diseases during the quarantine period. In other cases, to prevent the diseases spreading from animals that might be hiding an infectious agent without exhibiting clinical signs to resident animals, the quarantined animals will be tested for various diseases such as bovine tuberculosis, Johne's disease, brucellosis, leptospirosis, salmonellosis, campylobacteriosis, leucosis, bovine viral diarrhea (BVD), infectious bovine rhinotracheitis (IBR), trichomoniasis, neosporosis, ringworm, liver fluke, lungworm, digital dermatitis, and contagious mastitis pathogens (*Streptococcus agalactiae* and *Staphylococcus aureus*) [10, 14]. The testing of animals in the prophylactic quarantine is a valuable biosecurity tool when properly applied.

To prevent the bovine tuberculosis introduction, the biosecurity plan should take into consideration all possibilities of *Mycobacterium bovis* transmission. Cattle are the main reservoir and spread microbes through aerosols (adults) or manure (calves) to many domestic and wild mammalian species. Sheep, goats, pigs, horses, and dogs are spillover hosts and spread *M. bovis* spread microbes in various ways (respiratory, digestive, by bites, or scratches). After infection, badgers, brush-tail opossums, wild boars, deer, and other wildlife species become wildlife reservoirs (maintenance host). Humans are susceptible and contract the infection mainly by drinking raw milk and raw milk products. People with pulmonary or urogenital tuberculosis can retransmit the infection to cattle [22].

Calves are more susceptible and should be kept in a separate area to minimize their exposure to infectious agents [14]. Calves can carry many infectious diseases without clinical signs and positive results on the laboratory tests (e.g., Johne's disease). This risk can be reduced by purchasing calves only from herds officially certified as disease-free [1].

Because one of the most common ways of the BVD virus introduction in a free farm is via a pregnant heifer ("Trojan cow") carrying a persistently infected fetus, all calves from purchased cattle should be tested at birth to detect persistently infected animals with BVD virus [1, 9, 10]. Persistently infected animals are the main route of the BVDV spreading between herds because they cannot be detected by serological tests (immunotolerant calves), but excrete massive amounts of virus [1, 23]. The risk of farm contamination can be reduced by purchasing animals only from herds officially certified as BVDV-free. If the BVDV status in the herd of origin is unknown, then pregnant females should be isolated on arrival (the contact with any animal of breeding age must be restricted), tested for BVD antibody and BVD antigen, and released from isolation only if they are negative results at both tests or antibody positive, antigen-negative, calved, and the calf was tested negative or removed from the herd [1]. To prevent BVDV introduction into a free farm, the following risk factors should be considered: trade with live animals, embryo transfer and semen recipients, return of animals from animal exhibitions, direct contacts between cattle on pasture or over fences, density and activity of arthropod vectors, vaccination, and employee and visitors contact with animals [9, 24].

Sick and suspicious animals should be isolated in a specific area and always handled at the end. In the control of contagious mastitis, the latter are milked cows suspected of the disease [9].

Implementing effective biosecurity programs will bring long-term economic benefits. Dutch studies have shown that the main benefits of a closed dairy herd with good biosecurity are better fertility and lower slaughter rates. The USA comparative studies in Johne's disease-positive herds and Johne's disease-negative herds revealed an economic loss of almost US$ 100 per cow in positive herds. Spread of an infectious disease onto a farm can lead to large economic losses. An outbreak of BVD in an Australian farm with 320 milking cows caused losses of $AUD 144,700 [25].

Vaccination is another important biosecurity tool designed to protect resident cattle from infectious agents that could have been brought in by the newly purchased cows [26]. In dairy cattle, immunization mainly targets common infectious agents such as BVD virus, IBR virus, parainfluenza-3 (PI3) virus, bovine respiratory syncytial virus (BRSV), leptospirosis, and clostridial infections [27]. Vaccination programs should be established in collaboration with the herd veterinarian and adapted to the risk of the disease spreading on the farm, including infectious agents that evolve in the area [25, 28]. Vaccination should not be considered the primary or single biosecurity tool because no vaccine provides 100% immunity [26, 28].

Dairy herd vaccination programs are affected by various factors such as age and category of production, disease history, housing, type of vaccine (killed or modified live), and costs [28]. Vaccination programs are designed by age categories and are applied continuously to maximize herd immunity and minimize the spread of the infectious agent [27, 28].

Vaccination schedule for dairy heifers from birth to 6 months of age can be started with an oral modified live vaccine (MLV) for bovine rotavirus and bovine coronavirus given 30 minutes before the ingestion of colostrum to prevent the inactivation [28]. In the first hour of life, calves must receive 2.8 L of colostrum, and in the next 23 hours, the rest of 2.8 L [27]. Depending on the epidemiological situation, an intranasal vaccination of neonatal calves with respiratory vaccines (IBR/PI-3/BRSV) can be started at 3 days of age or older [28]. At 6 weeks old, dairy heifers can receive an injectable modified-live IBR/PI3/BRSV/BVD vaccine and a seven-way clostridial bacterin-toxoid [27]. The immunity of injectable vaccines is longer than the immunity of intranasal vaccines [28]. Following national and international regulations on brucellosis prophylaxis, at 4–6 months age replacement heifers should receive brucellosis RB51 vaccine. Also, depending on the epidemiological situation, calves can receive the appropriate vaccination for leptospirosis clostridial diseases and/or *Histophilus somnus*. At 6 months of age, heifers should be revaccinated with modified live IBR/PI3/BRSV/BVD virus vaccine, seven-way clostridial vaccine, and five-way leptospirosis bacterin [27, 28].

Pre-breeding heifers (10–12 months of age) should be revaccinated with killed or modified live IBR/PI3/BRSV/BVD virus vaccine, five-way leptospirosis bacterin, and seven- or eight-way clostridial bacterin-toxoid [28]. Optionally, it can be done with vibriosis bacterin [27].

Pre-calving heifers should be revaccinated 40–60 days before calving with killed IBR/PI3/BRSV/BVD virus vaccine, five-way leptospirosis bacterin, killed rotavirus and coronavirus vaccine, and *Escherichia coli* + *Clostridium perfringens* types C and D bacterin/toxoid. Three weeks before to calving, heifers should be revaccinated with killed rotavirus and coronavirus vaccine, and *Escherichia coli* + *Clostridium perfringens* types C and D bacterin/toxoid [27, 28]. Also, pre-calving heifers should be vaccinated with coliform mastitis bacterin [27].

Adult cows should be annually vaccinated, 40–60 days before calving for IBR, PI3, BRSV, and BVDV [27]. Depending on the history of diseases in the region and the associated epidemiological risks, the farm veterinarian should choose vaccines that immunologically protect dairy cows during the lactation period and the dry period for leptospirosis, vibriosis, *Rotavirus, Coronavirus, Clostridium perfringens* types C and D, and *Escherichia coli* mastitis. Types of vaccines recommended are killed or bacterin/toxoid and modified-live vaccines (MLV) [27, 28]. Adult dairy cattle should receive a booster vaccination at 3 weeks before calving with killed rotavirus and coronavirus vaccine and *Escherichia coli* + *Clostridium perfringens* types C and D bacterin/toxoid vaccine [27]. MLV vaccines should be used with prudence in pregnant cows and only after consultation with the veterinarian [28]. The annual vaccination for vibriosis should be performed in dairy herds where the artificial insemination is not practiced [27].

The annual vaccination of adult dairy cattle for calf scours (rotavirus and coronavirus, *Escherichia coli*, and *Clostridium perfringens* types C and D) should be considered in all herds with recent history as a part of the preventative management practices [27].

Mastitis is one of the most important diseases in dairy cows that affects the welfare, production, and duration of the economic life of the animals [29]. Economic losses are due to direct milk production losses (reduction of quantity, unsalable, or poor quality), culling or removal from the herd of animals with unsatisfactory treatment results, cost of veterinary care, cost of excessive use of antimicrobials and other medicines, and the risk of antibiotic resistance [30].

The main pathogens targeted by mastitis vaccines are *Staphylococcus aureus, Streptococcus agalactiae,* and *Escherichia coli* [29]. Reduction in the incidence and duration of intramammary infections can be obtained by applying the combination of vaccination with high milking hygiene procedures, treatment of clinical cases, segregation, and culling of known infected cows [29]. The following preventive measures were proved to have a positive result in the management of mastitis in dairy herds: the use of milkers' gloves, blanket use of dry-cow therapy, washing unclean udders, maintaining cows upright after milking, back-flushing of the milking cluster after milking an animal with clinical mastitis, and application of a treatment protocol [30] Also, to maximize the success of immunization, within 5 days of mastitis vaccines, dairy cows must not receive any other Gram-negative bacterin vaccines (e.g., *Escherichia coli*, *Salmonella* spp., *Pasteurella* spp., *Campylobacter* sp., and *Moraxella bovis*) [27].

To evaluate the effects of mastitis vaccines in dairy cows, the following monitoring parameters are most commonly used:

1. Clinical and subclinical mastitis incidence and severity

2. Somatic cell count

3. Serum and/or milk immunoglobulin G concentrations

4. Milk bacterial culture or *Staphylococcus aureus* count in milk

5. Milk production

6. Cure or cull rate [29]

Newly acquired dairy herd bulls should be 30–60 days in prophylactic quarantine and tested with negative results for persistent BVDV infection, brucellosis, and tuberculosis. Recommended vaccination schedule for dairy herd bulls is an annual vaccination at the breeding soundness examination with IBR/PI3/BVD killed vaccine, five-way leptospirosis bacterin, and vibriosis bacterin [27].

If there are animal species other than cattle, then the vaccination actions must take into account for these species as well. Farm dogs and cats should be vaccinated at least against rabies to protect humans and other animals [14].

Antibiotic overuse can be reduced by using a proper mixture of natural antibacterial peptides, biological response modifiers, prebiotics, probiotics, and correct development of the gut microbiome [31].

The limited use of bacterial culture and sensitivity testing by veterinarians are other causes of the persistence of the multidrug resistance (MDR) isolates in dairy farms. The findings of the last decades highlight the necessity of using antimicrobial susceptibility testing each time before prescribing an antibiotic [32].

4.2 Dead animal management

To reduce the risk of pathogens spreading in farm animals, dead animals should be disposed of in the shortest time. Depending on the national regulations and farm's possibilities, the disposal of carcasses can be done by a licensed dead stock collector, burial, or composting [14].

Studies designed to investigate what motivates and withholds farmers to implement biosecurity measures placed the carcass storage away from the stables on the second rank for feasibility, but with a lower score for efficacy [33].

Rendering trucks have a particular risk for farm biosecurity because they are at high risk for carrying animals killed by infectious diseases [26]. To prevent farm contamination, mortality pick-up should be located away from the stable and feed storage bin and silo [34].

5. Feed and water biosecurity

The biosecurity of feed and water must start from the source, respectively, from the fields where crops are grown and from the water capture source. Manure used as a natural fertilizer can contaminate the soil, crop, and water used for irrigation and groundwater sources [2]. The quality and potability of water should be tested regularly, and samples from each feedstuff batch or lot should be stored for possible laboratory analyses (e.g., bacteria, toxins, molds, and mycotoxins) until that batch is consumed without incidents [2, 10].

To reduce the risk of the diseases being introduced by contaminated feed, the dairy producer should record and monitor the manure application on its pastures and fields cultivated with feedstuffs [2]. The risk of a feed-related disease outbreak is increased when feedstuffs are purchased from multiple locations or the crops were fertilized with manure from other dairy farms [2, 10].

To prevent feedstuffs to be contaminated through fecal material and urine from rodents, birds, dogs, cats, and any wildlife, dairy farmers should design food storage areas in a way to be inaccessible (e.g., opened bags can be placed into containers with tight lids; barns can have welded wire fence) [2, 14].

The biosecurity plan of the dairy farm should include the frequency of storage areas cleaning, the way of feed bags storage off the floor on pallets, removing and disposing of the not consumed feed within 24 hours, rotation of feed inventory for the purpose to reduce the possible presence of detrimental organisms or toxins in stored feeds, and periodically checking of silos, bins, and bunks to detect and remove as soon as possible moldy or spoiled feedstuff [14].

Although not recommended, some cattle herds are still using surface water (e.g., lakes, ponds, and rivers) as a water source. Drinking water can be contaminated by animal carcasses (e.g., dead wild animals), manure from other livestock, bird droppings, urine and feces of wildlife, and human waste [2, 10, 14]. Water biosecurity programs should include several measures designed to prevent contamination with toxins and infectious agents such as restriction of the birds and wildlife access to farm water sources, filtration and chemical sterilization of water, and regular testing of water quality and potability [2]. Waterers should be cleaned once a week [14].

6. Manure biosecurity

In dairy farms, manure is the most problematic waste and should be treated as a biological risk material because it has a huge bacterial load [2]. Manure should be stored in an area inaccessible to cattle [14]. Contact with manure from infected cattle is the main means of spread for rotavirus, coronavirus, *Escherichia coli*, *Salmonellosis*, and Johne's disease to other receptive animals. Manure handling should prevent environmental contamination and should not violate the legislation in force [14].

Manure is rich in nutrients that could be recycled as fertilizer [35]. However, the use of this natural fertilizer should be done with caution to prevent contamination of crops, pastures, and groundwater sources [2]. *Salmonella* spp., *Escherichia coli*, *Listeria* spp., and *Mycobacterium avium* subsp. *paratuberculosis* can be killed by the process of manure composting but the process must be controlled before the use of compost in agriculture [2, 36, 37]. In the process of composting should not be used the manure from the hospital pen, where de infectious agents can be in a high concentration. Also, the temperature and microbial activity should be checked to confirm the complete sterilization [2, 14]. Also, manure can be recycled for bedding and to produce methane [2].

Manure biosecurity programs should include measures to prevent the manure equipment used to handle feed, the environment infestation with files and intestinal parasites (manure must be removed frequently to prevent the pest life cycles completion), manure run-off or transfer from adults to calves, and feed contamination by manure-covered wheels of farm vehicles [14].

Manure spreaders and slurry handling equipment are high-risk equipment and should be brought to the farm after proper cleaning or disinfection [1].

The manure cleaning of vehicles and equipment must be done in areas specially designed for this purpose, where water or disinfectants would not splash onto feed or into drinking water. Throughout the entire cleaning and disinfection process, the equipment will be inspected visually to dispel any suspicion of cross-contamination [2].

7. Conclusions

The development and implementation of biosecurity programs in dairy farms improve cattle health, welfare, and productivity. These programs must be monitored and evaluated continuously to identify new methods of control and new effective critical control points and to further improve the program to prevent the introduction and spread of infectious agents on the farm. The biosecurity program should be focused on the decision and adapted to the specific situations of each dairy farm. Many of the problems encountered can be prevented or minimized with the support of veterinary services. Staff and visitors should be trained on biosecurity measures applied on the farm.

Conflict of interest

The authors declare no conflict of interest.

Author details

Stelian Baraitareanu* and Livia Vidu
University of Agronomic Sciences and Veterinary Medicine of Bucharest, Romania

*Address all correspondence to: stelian.baraitareanu@gmail.com

References

[1] Sibley RJ. Biosecurity in the dairy herd. In: WCDS Advances in Dairy Technology. Vol. 26. Alberta, Canada: University of Alberta; 11-14 March 2014. pp. 59-74

[2] Villarroel A, Dargatz DA, Lane VM, McCluskey BJ, Salman MD. Suggested outline of potential critical control points for biosecurity and biocontainment on large dairy farms. Journal of the American Veterinary Medical Association. 2007;**230**(6):808-819. DOI: 10.2460/javma.230.6.808

[3] Hadorn DC, Stärk KD. Evaluation and optimization of surveillance systems for rare and emerging infectious diseases. Veterinary Research. 2008;**39**(6):57. DOI: 10.1051/vetres:2008033

[4] Dantzer R, Cohen S, Russo SJ, Dinan TG. Resilience and immunity. Brain, Behavior, and Immunity. 2018;**74**:28-42. DOI: 10.1016/j.bbi.2018.08.010

[5] Dargatz DA, Garry FB, Traub-Dargatz JL. An introduction to biosecurity of cattle operations. The Veterinary Clinics of North America. Food Animal Practice. 2002;**18**:1-5. DOI: 10.1016/s0749-0720(02)00002-6

[6] Shortall O, Green M, Brennan M, Wapenaar W, Kaler J. Exploring expert opinion on the practicality and effectiveness of biosecurity measures on dairy farms in the United Kingdom using choice modeling. Journal of Dairy Science. 2017;**100**(3):2225-2239. DOI: 10.3168/jds.2016-11435

[7] Nöremark M, Frössling J, Lewerin SS. A survey of visitors on Swedish livestock farms with reference to the spread of animal diseases. BMC Veterinary Research. 2013;**9**:184. DOI: 10.1186/1746-6148-9-184

[8] van Schaik G, Schukken YH, Nielen M, Dijkhuizen AA, Barkema HW, Benedictus G. Probability of and risk factors for introduction of infectious diseases into Dutch SPF dairy farms: A cohort study. Preventive Veterinary Medicine. 2002;**54**(3):279-289. DOI: 10.1016/s0167-5877(02)00004-1

[9] Wallace RL. Practical and sensible dairy farm biosecurity. In: Proceedings of the 6th Western Dairy Management Conference; 12-14 March 2003. Reno, NV: WDMC; 2003. pp. 201-206

[10] Villarroel A. Practical biosecurity on dairy farms. In: Oregon Veterinary Conference; 01 March 2007. Corvallis, OR: OVC; 2007. pp. 1-4. DOI: 10.13140/2.1.3657.7928

[11] Guidance. Disease Prevention for Livestock and Poultry Keepers. How to Prevent the Introduction and Spread of Animal and Bird Disease by Following Good Hygiene and Biosecurity Standards. 2015. Available from: https://www.gov.uk/guidance/disease-prevention-for-livestock-farmers#biosecurity-measures [Accessed: 02 May 2020]

[12] Caldow GL, Crawshaw M, Gunn GJ. Herd health security in the suckler herd. Cattle Practice. 1998;**6**:175-179

[13] Farm Biosecurity. Essentials. People, Vehicles & Equipment. Available from: https://www.farmbiosecurity.com.au/essentials-toolkit/people-vehicles-equipment/ [Accessed: 12 May 2020]

[14] Livestock Biosecurity. What Is it and Why Should I Care? Available from: https://dairy-cattle.extension.org/livestock-biosecurity/ [Accessed: 12 May 2020]

[15] Hoe FG, Ruegg PL. Opinions and practices of Wisconsin dairy producers about biosecurity and animal well-being. Journal of Dairy Science. 2006;**89**(6):2297-2308. DOI: 10.3168/jds.S0022-0302(06)72301-3

[16] Troutt HF, Galland J, Hyatt D, Rossiter C, Lein D, Brewer RL, et al. *Salmonella* and the market dairy cow: Transport contamination—Risk for farm biosecurity. The Bovine Practitioner. 2008;**42**:56-62

[17] BAMN/APHIS. An Introduction to Infectious Disease Control on Farms (Biosecurity). 2001. Available from: http://www.aphis.usda.gov/animal_health/nahms/dairy/downloads/bamn/BAMN01_IntroBiosecurity.pdf [Accessed: 12 May 2020]

[18] Pritchard G, Dennis I, Waddilove J. Biosecurity: Reducing disease risks to pig breeding herds. In Practice. 2005;**27**:230-237. DOI: 10.1136/inpract.27.5.230

[19] DEFRA Archive Website. Biosecurity Guidance to Prevent the Spread of Animal Diseases. 2003. Available from: https://webarchive.nationalarchives.gov.uk/20130402155521/http://archive.defra.gov.uk/foodfarm/farmanimal/diseases/documents/biosecurity_guidance.pdf [Accessed: 16 April 2020]

[20] Brennan ML, Christley RM. Biosecurity on cattle farms: A study in north-west England. PLoS One. 2012;**7**(1):e28139. DOI: 10.1371/journal.pone.0028139

[21] Technical Information. Buildings. Housing. Building Biosecurity. 2020. Available from: https://dairy.ahdb.org.uk/technical-information/buildings/housing/building-biosecurity/#.XsU6wkQzaHs [Accessed: 16 April 2020]

[22] Baraitareanu S. Infectious Diseases, Preventive Medicine and Clinical Lectures on Species 2. Course Manual. Printech: Bucharest; 2020. p. 180

[23] Baraitareanu S, Vidu L. The preventive medicine of bovine viral diarrhoea-mucosal disease in dairy farms: A review. Revista Romana de Medicina Veterinara. 2019;**29**(2):61-64

[24] Lindberg AL, Alenius S. Principles for eradication of bovine viral diarrhoea virus (BVDV) infections in cattle populations. Veterinary Microbiology. 1999;**64**(2-3):197-222. DOI: 10.1016/s0378-1135(98)00270-3

[25] Cockcroft PD, editor. Bovine Medicine. 3rd ed. Chichester: Wiley; 2015. p. 644

[26] Sanderson M. Biosecurity for cow-calf enterprises. In: Anderson DE, Rings DM, editors. Food Animal Practice. 5th ed. Philadelphia: Saunders; 2009. pp. 594-599. DOI: 10.1016/b978-141603591-6.10113-7

[27] Waldner DN, Kirkpatrick J, Lehenbauer TW. Recommended Vaccination Schedules for a Comprehensive Dairy Herd Health Program. 2017. Available from: https://extension.okstate.edu/fact-sheets/recommended-vaccination-schedules-for-a-comprehensive-dairy-herd-health-program.html [Accessed: 16 April 2020]

[28] Dewell G, Gorden P, Breuer R. Iowa State University Extension and Outreach. Dairy Cattle Vaccination Programs. 2016. Available from: https://store.extension.iastate.edu/Product/da3088-pdf [Accessed: 16 April 2020]

[29] Ismail ZB. Mastitis vaccines in dairy cows: Recent developments and recommendations of application. Veterinary World. 2017;**10**(9):1057-1062. DOI: 10.14202/vetworld.2017.1057-1062

[30] Hogeveen H, Huijps K, Lam TJ. Economic aspects of mastitis: New developments. New Zealand Veterinary Journal. 2011;**59**(1):16-23. DOI: 10.1080/00480169.2011.547165

[31] Trevisi E, Zecconi A, Cogrossi S, Razzuoli E, Grossi P, Amadori M. Strategies for reduced antibiotic usage

in dairy cattle farms. Research in Veterinary Science. 2014;**96**(2):229-233. DOI: 10.1016/j.rvsc.2014.01.001

[32] Lawrence KE, Wakeford L, Toombs-Ruane LJ, MacLachlan C, Pfeffer H, Gibson IR, et al. Bacterial isolates, antimicrobial susceptibility and multidrug resistance in cultures from samples collected from beef and pre-production dairy cattle in New Zealand (2003-2016). New Zealand Veterinary Journal. 2019;**67**(4):180-187. DOI: 10.1080/00480169.2019.1605943

[33] Damiaans B, Sarrazin S, Heremans E, Dewulf J. Perception, motivators and obstacles of biosecurity in cattle production. Vlaams Diergeneeskundig Tijdschrift. 2018;**87**(3):150-163

[34] Payne M. CDQAP Ruminations: Dairy Biosecurity & Your Bottom Line. 2015. Available from: http://cdrf.org/2015/05/14/cdqap-ruminations-dairy-biosecurity-your-bottom-line/ [Accessed: 10 May 2020]

[35] Hogan JS, Bogacz VL, Thompson LM, Romig S, Schoenberger PS, Weiss WP, et al. Bacterial counts associated with sawdust and recycled manure bedding treated with commercial conditioners. Journal of Dairy Science. 1999;**82**(8):1690-1695. DOI: 10.3168/jds.S0022-0302(99)75398-1

[36] Lung AJ, Lin CM, Kim JM, Marshall MR, Nordstedt R, Thompson NP, et al. Destruction of *Escherichia coli* O157:H7 and *Salmonella enteritidis* in cow manure composting. Journal of Food Protection. 2001;**64**(9):1309-1314. DOI: 10.4315/0362-028x-64.9

[37] Grewal SK, Rajeev S, Sreevatsan S, Michel FC Jr. Persistence of *Mycobacterium avium* subsp. *paratuberculosis* and other zoonotic pathogens during simulated composting, manure packing, and liquid storage of dairy manure. Applied and Environmental Microbiology. 2006;**72**(1):565-574. DOI: 10.1128/AEM.72.1.565-574.2006

Chapter 7

Antimicrobial Resistance with Special Emphasis on Pathogens in Agriculture

Nitya Meenakshi Raman, Murugesh Easwaran, Rashmi Kaul, Jyotsna Bharti, Khaled Fathy Abdel Motelb and Tanushri Kaul

Abstract

Antibiotics have been used globally to manage the bacterial plant diseases irrespective of the expense involved. Although plant pathogenesis by bacteria is far lower than fungal counterparts, disrupted monitoring and surveillance for drug resistance with respect to human health raise serious concerns. The resistance derived by the plant as the host by the antibiotics used for many generations has now posed as a problem in phyto-systems. Although we currently lack the molecular understanding of the pathogens rendering antibiotic resistance to plants, robust resistance management strategies are critical to ensure management of critically important diseases that specifically target crops of high value and/or global agrarian importance. This chapter discusses evolution of plant-pathogenic bacteria, application of antibiotics and its repercussions on the microbiome of plant agricultural systems, and sustainable crop disease management by genetic engineering.

Keywords: agriculture, bacteria, fruit, genetic engineering, host, molecular biology

1. Introduction

Antibiotic resistance most commonly evolves in bacteria either through mutation of a target site protein, through the acquisition of an antibiotic-resistant gene that confers resistance through efflux or inactivation of the antibiotic, or through synthesis of a new target protein that is insensitive to the antibiotic [1]. An extensive body of knowledge has been gained from studies of antibiotic resistance in human pathogens and in animal agriculture. The ability of bacterial pathogens to acquire antibiotic-resistant genes and to assemble them into blocks of transferable DNA encoding multiple antibiotic-resistant genes has resulted in significant issues that affect successful treatment interventions targeting some specific human infections. The current global antibiotic resistance crisis in bacterial populations has been fuelled by basic processes in microbial ecology and population dynamics, engendering a rapid evolutionary response to the global deployment of antibiotics by humans in the millions of kilograms per year. What was not anticipated when antibiotics were discovered and introduced into clinical medicine is that antibiotic-resistant genes pre-existed in bacterial populations [2–4]. Furthermore, the extent to which antibiotic-resistant genes could be transferred between bacteria, and even between phylogenetically distinct bacteria, was not understood 70 years ago but is becoming more apparent through a number of elegant studies identifying the microbial antibiotic resistome. The collection of all known antibiotic-resistant genes in the full-microbial pan-genome is defined as the antibiotic resistome [5].

2. Use of antibiotics in agriculture

Effective management of bacterial plant diseases is difficult and is exacerbated by factors such as the large size of bacterial pathogen populations on susceptible plant hosts and the few available bactericides. In the absence of durable and robust host disease resistance, antibiotics have represented the best option for bacterial disease control in many pathosystems because these materials provide the most efficacious means of reducing bacterial population size and limiting disease outbreaks. Although many new types of antibiotics were rapidly tested and then deployed in animal agriculture starting in the 1950s, antibiotic use for plant disease control was tempered by several factors, including lack of efficacy at lower doses, phytotoxicity problems at higher doses, and expense compared to other existing methods of disease control. Thus, although penicillin, streptomycin, aureomycin, chloramphenicol, and oxytetracycline were tested for plant disease control in the late 1940s [6, 7], only streptomycin and oxytetracycline were ultimately deployed in plant agriculture and only in specific disease pathosystems. Streptomycin is the main antibiotic currently in use for plant disease control around the world, targeting pathogens such as *Erwinia amylovora*, which causes fire blight of apple and pear; *Pseudomonas syringae*, which causes flower and fruit infection of apple and pear trees; and *Xanthomonas campestris*, which causes bacterial spot of tomato and pepper [8]. Oxytetracycline has been used as the primary antibiotic in specific disease control situations, including the control of *Xanthomonas arboricola* pv. *pruni*, the causal agent of bacterial spot of peach and nectarine [8]. In addition, oxytetracycline has been used as a secondary antibiotic for fire blight management in the United States, most prominently in situations in which streptomycin resistance has become a problem [9, 10].

The problem of antibiotic resistance is not limited to the Indian subcontinent only, but is a global problem. To date, no known method is available to reverse antibiotic resistance in bacteria. The discovery and development of the antibiotic penicillin during the 1900s gave a certain hope to medical science, but this antibiotic soon became ineffective against most of the susceptible bacteria. The antibiotic resistance in bacteria is generally a natural phenomenon for adaptation to antimicrobial agents. Once bacteria become resistant to some antibiotic, they pass on this characteristic to their progeny through horizontal or vertical transfer. The indiscriminate and irrational use of antibiotics these days has led to the evolution of new resistant strains of bacteria that are somewhat more lethal than the parent strain. More recently, in 2016, a Section 18 emergency exemption was granted by the US Environmental Protection Agency for the use of streptomycin and oxytetracycline on citrus trees in Florida for management of citrus Huanglongbing (HLB) disease [11–13]. Regarding other antibiotics, gentamicin has been used in Mexico for fire blight control and in Chile, Mexico, and Central American countries for vegetable disease control, while oxolinic acid (OA) has been used only in Israel for fire blight management [14, 15]. Lastly, kasugamycin is used in Japan and other Asian countries to control the fungal disease rice blast and bacterial seedling diseases of rice [16] and has recently been registered for use in the United States and Canada for managing fire blight [17]. Concerns regarding the use of antibiotics in plant disease control and potential impacts on human health have led to the banning of antibiotic use by the European Union. However, streptomycin is still utilized for fire blight management in Austria, Germany, and Switzerland under strict control parameters.

3. Evolution of plant-pathogenic bacteria

3.1 Resistance to streptomycin

The lack of effective bactericide alternatives in several plant disease systems has resulted in a decade-long dependence or overdependence on streptomycin. As streptomycin has been used the longest, over the largest geographic area, and for treatment of the largest variety of crops, streptomycin resistance is relatively widespread among plant-pathogenic bacteria. Although the first streptomycin-resistant (SmR) plant-pathogenic bacteria detected were strains of *E. amylovora* harboring a chromosomal resistance mutation, the majority of SmR plant pathogens encode the transmissible SmR transposon Tn5393 [8]. Tn5393 is a Tn3-type transposon originally isolated from *E. amylovora* that harbors *strAB*, a tandem resistance gene pair that confers streptomycin resistance through covalent modification of the streptomycin molecule [18]. The Tn5393 transposon is composed of genes required for the transposition process (*tnpA* and *tnpR*), a central site that contains outwardly directed promoters for expression of both *tnpA* and *tnpR* as well as the *strAB SmR* genes. Expression of the *strAB* genes from Tn5393 in *E. amylovora* is driven by a promoter present in the 3 prime end of the insertion sequence IS1133 that is inserted directly upstream of the *strA* gene [19]. Two closely related variants of Tn5393 have also been found in plant pathogens: Tn5393a, an element that does not contain IS1133, has been detected in *P. syringae* and in a group of *E. amylovora* strains from California exhibiting a moderate level of resistance, and Tn5393b, an element that does not contain IS1133 but instead contains an insertion of IS6100 within the *tnpR* gene, has been characterized in *X. campestris* [19, 20].

There are two other reports of additional genetic mechanisms of streptomycin resistance in plant pathogens; these include the occurrence of the small, nonconjugative but mobilizable broad-host-range plasmid RSF1010 in some strains of *E. amylovora* isolated in California [21]. This observation carries further significance because RSF1010 has been distributed globally among a number of bacterial genera and also occurs in some human-pathogenic bacteria [22]. A recent report detailing an analysis of streptomycin-resistant *X. oryzae* subsp. *oryzae* from China indicated that four strains harbored the *aadA1* gene associated with class 1 integron sequences [23]. This observation is significant because of the importance of integrons in both the transfer of antibiotic resistance in human and animal pathogens and the accumulation of antibiotic resistance genes within one multiresistance element. To date, streptomycin resistance mediated by Tn5393 or the closely related variants has been reported in *E. amylovora*, *P. syringae*, and *X. campestris* isolated from North and South America and Asia [19, 20, 24–30]. The location of essentially the same genetic element in different genera of plant pathogens isolated from distinct crop hosts and from different continents is confirmatory evidence of the role of horizontal gene transfer (HGT) in the dissemination of antibiotic resistance in these pathosystems.

The source of Tn5393 to the plant pathogens was likely not from the antibiotic preparations themselves as a study of 18 available agricultural streptomycin formulations revealed no contamination with the *strA SmR* gene [31]. Instead, the acquisition of Tn5393 by bacterial plant pathogens was likely from commensal co-occurring epiphytic bacteria via HGT. For example, Tn5393 was thought to have been acquired by *E. amylovora* on the plasmid pEa34 from *Pantoea agglomerans*, a common orchard epiphyte [18]. The transfer event most likely occurred on the apple flower stigma, a surface where *E. amylovora* grows to high population densities and where *Pantoea agglomerans* can also grow. *Pseudomonas syringae* and *X. campestris* pv. *vesicatoria* both have epiphytic phases where the pathogens grow on leaf surfaces, providing opportunities for HGT with other epiphytes. It should be noted that high-level streptomycin resistance, conferred by a spontaneous mutation within the *rpsL* gene that encodes the ribosomal target protein for streptomycin, does occur in some populations of *E. amylovora*, particularly within populations from the western United States as well as in a small number of strains isolated in Michigan and New Zealand [32, 33]. The minimal inhibitory concentration (MIC) of streptomycin in these highly resistant spontaneous mutants is greater than 4096 μg/mL [32]. In contrast, SmR strains of *E. amylovora* harboring Tn5393 exhibit MICs of streptomycin ranging from 512 to 1024 μg/mL [32]. Streptomycin solutions used for fire blight management are typically applied at 100 μg/mL; thus, it is unclear whether the increased level of resistance exhibited by the spontaneous mutants provides a survival advantage in streptomycin-treated orchards.

3.2 Resistance to tetracyclin

Tetracycline resistance has been reported in a few plant-pathogenic bacteria, including *P. syringae* [34, 35] and *Agrobacterium tumefaciens* [36]. Other studies have reported on sensitivity; for example, in one study, 138 strains of *E. amylovora* from the Pacific Northwest, USA, were all determined to be sensitive to oxytetracycline [37]. Although there are few reports of resistance, multiple tetracycline resistance genes homologous to *tetA* and *tetM* are present within the genomes of many different plant-pathogenic bacteria. Efflux pump proteins that belong to the same protein family as TetA have been identified in *Ralstonia solanacearum*; *Erwinia piriflorinigrans*; multiple *Xanthomonas* species, including *Xanthomonas citri*, *Xanthomonas phaseoli*, *Xanthomonas perforans, and X. campestris*; multiple *Pseudomonas* species, including *P. syringae*, *Pseudomonas aeruginosa*, and nonpathogenic *Pseudomonas putida* and *Pseudomonas fluorescens*. However, even though putative tetracycline-resistant proteins have been annotated in the NCBI database for plant-pathogenic bacteria such as *Erwinia*, *Pseudomonas*, *Xanthomonas*, *Agrobacterium*, and *Ralstonia*, their function in tetracycline resistance remains to be characterized.

3.3 Resistance to oxolinic acid and kasugamycin

There are a few reports documenting resistance to other antibiotics used in plant disease management. OA was introduced in 1997 for fire blight management in Israel as a replacement for streptomycin, and OA resistance in *E. amylovora* was first detected in 1999 [38] and expanded in range by 2001 [39]. However, populations of OA-resistant *E. amylovora* fluctuated, with OA-resistant strains becoming undetectable in orchards where they previously occurred. Laboratory analyses of OA-resistant strains suggested that these strains were reduced in fitness compared to OA-sensitive strains [40]. Analysis of OA-resistant strains of *Burkholderia glumae* also showed that the strains were reduced in fitness, as these strains could not survive in rice paddy fields [41]. Kasugamycin was discovered in Japan and has been used since the 1960s in Asia for the control of rice blast caused by the fungus *Magnaporthe grisea* and for the control of bacterial grain and seedling rots of rice. This antibiotic has also been used to control diseases of sugar beet, kiwi, and Japanese apricot in at least 30 countries [42]. More recently, kasugamycin has been utilized for management of the blossom blight phase of fire blight disease in Canada and the United States. Resistance to kasugamycin was reported for two bacterial rice pathogens in Japan, *Acidovorax avenae* subsp. *avenae* and *Burkholderia glumae* [43, 44]. Kasugamycin resistance in *A.*

avenae subsp. *avenae* and *B. glumae* was conferred by a novel aac(2)-IIa acetyltransferase gene located within an IncP genomic island and likely acquired by HGT [45]. A promoter mutation that resulted in a fourfold increase in expression of the aac(2)-IIa gene was found to confer an increased level of kasugamycin resistance in strain 83 of *A. avenae* subsp. *avenae* [46]. Kasugamycin resistance has not been reported in *E. amylovora*; one study assessing the potential for spontaneous resistance revealed that a two-step mutational process was required and that spontaneous kasugamycin resistant mutants were substantially reduced in fitness [17].

4. Application of antibiotics and its repercussions on the microbiome of plant agricultural systems

All of the antibiotics applied to trees in orchard systems using conventional air blast spraying systems does not reach the desired target; thus, the effects of antibiotic usage are potentially more complex than simply studying effects on the target pathogen and commensals co-located in the target plant habitat. Antibiotics reaching the target sites in the tree canopy impact the phyllosphere microbiome and flower microbiomes if applied during the bloom phase. Insects feeding within the tree canopy could also ingest the antibiotic, which could impact the insect gut microbiota. A portion of the antibiotic spray applied to trees will not reach the target because of spray drift or could be lost by runoff during spraying or runoff owing to rain events. It has been estimated that as much as 44–71% of spray solutions applied by air blast sprayers is lost into the environment [47]. Whether it hits the target or not, once the antibiotic solution has been released into the environment, the material is negatively affected by environmental parameters, including rainfall, sunlight (visible and ultraviolet radiation), and temperature, and other specific aspects of the plant leaf environment that may affect adsorption. For example, oxytetracycline residues are lost relatively rapidly from peach leaf surfaces because of weather parameters [48]. Any antibiotic lost from the tree target by spray drift may land on other plant surfaces, such as the leaves of grasses or weeds, and thus impact the microbes inhabiting the phyllosphere of those plants. There is also the possibility of drift offsite to nontarget plants, and insect or animal may feed on the nontarget plants and potentially consume the antibiotic, which could impact the gut microflora of these animals. We are aware of one study in which the percentage of streptomycin-resistant *E. coli* isolates from feces of sheep feeding in a pasture that was sprayed with streptomycin was shown to increase (from 14.7 to 39.9% compared to 15.8 to 22.3% in a control group) [49]. However, this study did not simulate actual conditions in commercial orchards as the streptomycin solution was sprayed directly onto the pasture grass and sheep were grazed in the pasture for 12 h immediately following application. Neither of these situations occurs in commercial orchards.

Two studies have been published examining the effect of antibiotic application in apple orchards on phyllosphere bacteria. In one study using both culture-based and culture-independent approaches, Yashiro and McManus [50] examined phyllosphere bacteria from apple orchards that either had received streptomycin applications in spring for fire blight management for up to 10 previous years or had not been sprayed. The percentage of culturable isolate resistant to streptomycin was actually larger from the non-sprayed orchards. An examination of community structure using 16S rRNA clone libraries indicated that streptomycin treatment did not have long-term effects on the diversity or phylogenetic composition of the phyllosphere bacterial community in the examined apple orchards [50]. A separate cultural study evaluated the effect of weekly applications of streptomycin (for 0, 3, 5, and 10 weeks) beginning at 80% bloom on specific components of the phyllosphere community. Testing of orchard epiphytes for streptomycin resistance indicated that 76.2, 94.5, 95.5, and 98.5% of the bacterial isolates were resistant to streptomycin on trees receiving 0, 3, 5, and 10 applications within one season, respectively [51]. Further microbiome studies have also been conducted examining the effect of antibiotic usage on soil microbiomes in apple orchards. For example, Shade et al. [52] determined that streptomycin application to apple trees did not result in any observable difference in soil bacterial communities (soil collected beneath trees 8–9 days after streptomycin application). The authors concluded that application of the antibiotic had minimal impact on nontarget bacterial communities [52]. A second microbiome study of apple orchard soil collected 14 days after streptomycin application also failed to detect any influence of the antibiotic on the soil bacterial community [53].

The microbiome studies detailed above have provided information that show limited impacts of antibiotics on the selection of antibiotic resistance at a period of time after application. However, there are no published studies to date assessing the resistome of crop plants and in particular the resistome of crop plants that have been treated with antibiotics. Interestingly, the application of struvite ($MgNH_4PO_4{\cdot}6H_2O$), which has been used as a plant fertilizer, alters the antibiotic resistome in the soil, rhizosphere, and phyllosphere [54]. This might have resulted from the fact that struvite usually contains ARGs, antibiotic-resistant bacteria, and antibiotic residues [50]. The need for knowledge of the antibiotic resistome in plant agricultural systems and especially in plant agricultural systems in which antibiotics are applied is critically important because we need to understand whether the use of antibiotics in plant agriculture has the potential to select ARGs that could impact human health. This issue regarding potential impacts to human health is highly significant, with current implications for the use of antibiotics in animal agriculture [55, 56]. Identification of particular ARGs, and the organisms harboring these genes, is important for risk assessments of pathogen acquisition of resistance based on close phylogenetic relationships with coinhabiting antibiotic-resistant commensals. If ARGs of importance in clinical medicine are identified in the resistome of plants sprayed with antibiotics, it is critical to determine whether their frequency and/or bacterial host range changes based on antibiotic exposure.

5. Knowledge gaps in plant-pathogen system

One of the gaps involved in the understanding of the host-plant-environment interaction is the attributes involved with respect to the change in climatic conditions. Changes brought about by the pathogen populations to the host are influenced by cultural practices, control methods, introduction of new cultivars or varieties, and climatic variability in equal measure. A majority of these studies are often hindered due to the difficulty in obtaining the information or evidence with respect to the presence of the pathogen throughout the said period, genetic composition and its associated changes before and after interaction with the host, climatic requirements for the host and pathogen during the said period and arrive at a convincing trend without background noise with respect to the disease pattern.

Similar to the pathogen-human interaction, the challenge and attack by pathogenic organisms are halted by the defense mechanisms of the plants. This mechanism is often trespassed by the evolution and emergence of newly faced pathogens that have evolved in response to evolution or agricultural practices and colonization strategies in native communities with no prior evolutionary history [57–59]. It is well-known that the ecosystem, frequency, and evolution of both host and pathogens are largely dependent on catastrophic outbreaks that have a direct involvement of the human population. Added to this is the development of a new species, migration of humans, speciation, susceptibility of the plants, divergence, and climate change [58]. With a positive association between the emergence of new pathogens and extinction of crop production being rendered by many researchers, understanding and identification of emerging pathogens is a necessary strategy to counter them [60, 61].

Understanding the emergence of new pathogens has largely been a challenge for scientists as the host-pathogen interaction is a complex process. Global distribution and diversity of plant pathogens is also dependent on trade, human migration, plant ecosystem, and distribution of plant-based products. An additional indirect way to gauge pathogens and their associated effects is the elucidation of migration pathways [62]. The ever-increasing investment by the researchers in analyzing genome sequences has revealed another world of improvement in understanding the adaptability of pathogens to plant disease [63–65], and any changes in pattern of pathogenicity may thus arise. Horizontal gene transfer and interspecific hybridization have been the two mechanisms that have been comprehensively reviewed [58, 63, 66–69]. Along with strategies such as population genomics study for development of improved disease management, awareness of agricultural heterogeneity and management or restriction of movement of plant materials aids have also been integrated. Further a cumulative effort by plant epidemiologists, ecologists, pathologists, and academic researchers facilitates successful management of emerging phytopathogens.

6. Sustainable crop disease management by genetic engineering (GE)

In addition to a plethora of published GE strategies, ongoing research, and the wide expansion of genetic resources, conceivable applications are gaining momentum [70] that invests prospective for future generations. The dynamics of the adaptation of pathogen toward the host can be invested by GE strategies due to its selective efficacy against a group or particular target pathogens. Such a targeted advantage minimizes health concerns at the consumers' end with no risk of nontarget biota in an agrarian ecosystem. Some of the processes that occur naturally have also been undertaken in GE processes (**Table 1**). Although the futuristic potential of GE strategies with controlled disease conditions in the subsequent host generations is questionable in the present day, it is demonstrated that GE strategies that were initiated as a proof of concept are now well-established and have been marketed as commercially viable varieties.

6.1 Boosting plant recognition of infection

Similar to a human system, plants also trigger defense molecules on recognizing particular molecules of invading pathogens generally referred to as pathogen-associated molecular patterns (PAMPs; [71–73]) that illicit a PAMP-triggered immunity. Although PAMP receptor molecules differ among plant species, genes that encode PAMP receptor can be transformed into other crops for triggering immunity [73]. Such a method of transformation does not introduce a novel defense mechanism but rather introduces a receptor that helps the transformed plant recognize infection making it independently counter the infection by its natural immune system [74–77].

Plant species	Disease	Pathogen species	Pathogen class	Gene product	Reference
Arabidopsis	Crown gall disease	*Agrobacterium tumefaciens*	Bacteria	Arabinogalactan protein	[78, 79]
	Crown gall disease	*Agrobacterium tumefaciens*	Bacteria	Mannan synthase	
	Root-knot nematode	*Meloidogyne incognita*	Nematode	Kelch repeat protein	[80, 81]
	Powdery mildew	*Erysiphe orontii*	Fungus	Receptor-like kinase	[82]
	Root-cyst nematode	*Heterodera schachtii*	Nematode	Ethylene response	[83, 84]
	Bacterial speck	*Pseudomonas syringae*	Biotrophic bacteria	Lectin receptor kinase	[22]
	Gray mold/rot; leaf spot	*Alternaria brassicicola; Botrytis cinerea*	Necrotropic fungus	Expansin	[85]
	Powdery mildew	*Golovinomyces orontii*	Biotrophic fungus	Membrane-attached protein	[86]
	Downy Mildew	*Hyaloperonospora arabidopsidis*	Biotrophic oomycete	ADP ribosylation factor—GTPase activating factor	[87]
	Bacterial wilt	*Ralstonia solanacearum*	Biotrophic bacteria	MAPkinase phosphatase	[88]
	Aphid	*Myzus persicae*	Insects	Fatty acid desaturase	[89]
Maize	Southern corn leaf blight	*Bipolaris maydis/Cochliobolus heterostrophus*	Necrotrophic fungus	Mitochondrial transmembrane protein	[90]
	Powdery mildew	*Blumeria graminis*	Biotrophic fungus	Long-chain aldehyde synthesis	[91]
Tomato	Gray mold/rot	*Botrytis cinerea*	Necrotrophic fungus	Polygalacturonase and expansin	[92]
	Soft rot, gray mold/rot	*Botrytis cinerea, Erwinia chrysanthemi*	Fungus, bacteria	ABA aldehyde oxidase	[93]
	Powdery mildew	*Leveillula taurica*	Biotrophic fungus	Membrane-anchored protein	[94]
	Aphid	*Macrosiphum euphorbiae*	Insects	Fatty acid desaturase	[95, 96]
	Fusarium wilt	*Fusarium oxysporum*	Hemibiotrophic fungus	Lipid transfer protein	[97]
Rice	Bacterial blight	*Xanthomonas oryzae*	Bacteria	MAPKKK	[98]
	Blight rot	*Burkholderia glumae*	Bacteria	MAP kinase	[99]
	Rice blast	*Magnaporthe oryzae*	Hemibiotrophic fungus	Transcription factor WRKY	[100, 101]
	Leaf blight	*Xanthomonas oryzae*	Bacteria	Stearoyl-ACP desaturase	[102]

Table 1.
Genes and their contributions to the plant-pathogen interaction studies.

6.2 Mining R genes

An intracellular receptor protein (R-protein) is produced as a mechanism of effector-triggered susceptibility which is banked on by a model of disease resistance [72, 103]. This protein is specifically detected in the presence or when an activity of a pathogen effectors is triggered resulting in effector-triggered defense [103]. However, these effectors may modify or change the defense response in the host in response to a new effector produced by the pathogen. With this production of specific R genes with respect to the pathogen effector, pools of resistance genes evolved can be made useful in breeding crops for disease resistance by producing cisgenics [104]. Exceptional efforts by conventional introgression of cisgenes undertaken in crops such as apple, banana, grape, and potato have established it to be labor intensive and time consuming [73, 104]. GE strategies offer a major advantage not only by making it easier and faster but also evading linkage drag [50, 74]. Further introgression of R genes can be made feasible between unrelated plant species among monocots and dicots [77, 105–108]. The tendency of the pathogen to overcome the resistance rendered by R genes can be circumvented by mining R genes from unrelated species by integrating GE strategies and breeding [109, 110].

6.3 Upregulating defense pathways

The activity of defense can be boosted by targeting molecules such as reactive oxygen species, pathogenesis-related genes involved in defense regulation, signaling, and associated processes activating acquired resistance. Such measures were profited to a great extent in enhancing resistance to diseases such as citrus greening and pathogens such as *Rhizoctonia solani* and *Magnaporthe oryzae* that utilizes the plant's own natural immune system without the introduction of new or novel metabolic pathways [111, 112].

6.4 Disarming host susceptibility genes

Some important genes that facilitate normal physiology in plants have been observed to be involved in facilitating pathogen colonization and infection. Changes induced in such susceptibility genes is an efficient strategy for disease resistance [113]. Disarming susceptibility genes may alter the pathosystems and many host factors that contribute to compatibility between the pathogen and host. Gaining a new function to replace the lost host factor is not a likely by the pathogen to overcome the activity of a disarmed susceptibility gene; therefore, this strategy does not leave any exogenous DNA [113].

6.5 Silencing essential pathogen genes

RNA interference is elicited in plants to silence genes that render pathogenicity by using genetic constructs with identical sequence of dsRNA. Such efforts directly trigger posttranscriptional gene silencing of the natural disease process. Such a process of silencing does not generate a biochemical pathway or produce a novel protein. Integrating the need of the hour with the potential of the strategy of RNA silencing proved profitable for the papaya industry in Hawaii [114, 115]. Such applications are observed in cases where severe strains of the virus can be reduced in case of an infection by a mild strain. Implementing a natural phenomenon for cross-protection as a means to manage disease conditions has practical drawbacks. These drawbacks were controlled by feeding insects with dsRNA constructs that can trigger RNAi [116, 117].

7. Engineering CRISPR/Cas immune system

Clustered regularly interspaced short palindromic repeats has been identified to be a prokaryotic defense system that combines with its associated proteins (Cas) to render an endonuclease activity that cuts the invading DNA at a particular target of interest. This specificity is determined by the sequence of DNA that matches the sequence of the RNA guide strand associated with the Cas protein. Some studies have engineered a Cas9/gRNA that targets the replicating DNA of Gemini virus that leads to agrarian crisis in tropical and subtropical climates [118–120]. Significant resistance to host can be achieved against a DNA virus by a targeted sequence-specific engineered complex of Cas9/gRNA, although the results are meant to be reproducible [121]. Long-term utilization of this strategy against a variety of genetic elements that hamper the host such as viruses can be successfully targeted [122–126].

Genome editing, brought about by *Agrobacterium*-mediated transformation or biolistic methods, gives way to a wide range of possibilities for genetic changes. Targeted modifications, specific mutagenesis, and /or modest changes can be brought about by targeting existing genes in live cells. By using CRISPR/Cas9, it is possible to create a non-transgenic gene edit that can be introgressed by conventional breeding and can yield a change that cannot be distinguished from a mutation [127]. Another application of CRISPR is that the genome editing is HDR-based that allows editing a gene from the crop's natural pool giving rise to cisgenic lines that can achieve outcomes stabilized by conventional breeding. HDR-based genome editing strategies also helps add a specific gene from an evolutionary distant organism therefore making the regulatory scrutiny mandatory similar to that of transgenics [128, 129]. Various research groups have validated CRISPR/Cas9 techniques to be straightforward, low cost, and efficient, but the accessibility of the applications of genome editing is largely dependant on democratizing genome editing, nonprofit organizations, and governmental regulations.

8. Conclusion

While recognizing the important benefits GE technologies offer, larger considerations merit attention, especially questions of public acceptability and of whether there are any long-term ecological risks different from those posed by conventional breeding. In considering such issues, it is important to remember that, not only do diverse GE strategies exist, but diverse GE manipulations are possible, ranging from very modest, targeted mutagenesis, through cisgenics and intragenics, to insertion of transgenes from other crops, from other (non-crop) plants, and from evolutionarily distant organisms. Thus, in considering socioeconomic and cultural perspectives of GE, it is important to bear in mind this diversity of strategies and applications: GE crops can differ markedly from one another. A useful GE construct may target one or a few pathogens of particular importance, but other breeding techniques still is important for tackling disease problems not targeted by available GE traits. Thus, GE should be understood, not as the best approach to addressing sustainability challenges, but as a suite of tools that capitalizes on the knowledge that biologists gain through our ongoing study of Nature. GE simply expands the breeding "toolbox," providing options to consider on a case-by-case basis for enhancing the sustainability of crop disease management.

DOI: http://dx.doi.org/10.5772/intechopen.88440

Author details

Nitya Meenakshi Raman, Murugesh Easwaran, Rashmi Kaul, Jyotsna Bharti, Khaled Fathy Abdel Motelb and Tanushri Kaul*
Nutritional Improvement of Crops, Plant Biology Division, International Center for Genetic Engineering and Biotechnology (ICGEB), New Delhi, India

*Address all correspondence to: kaultanushri3@gmail.com

References

[1] Davies J, Davies D. Origins and evolution of antibiotic resistance. Microbiology and Molecular Biology Reviews. 2010;**74**(3):417-433

[2] Bhullar K, Waglechner N, Pawlowski A, Koteva K, Banks ED, Johnston MD, et al. Antibiotic resistance is prevalent in an isolated cave microbiome. PLoS One. 2012;7(4):e34953

[3] Knapp CW, Dolfing J, Ehlert PA, Graham DW. Evidence of increasing antibiotic resistance gene abundances in archived soils since 1940. Environmental Science & Technology. 2009;**44**(2):580-587

[4] Perry J, Waglechner N, Wright G. The prehistory of antibiotic resistance. Cold Spring Harbor Perspectives in Medicine. 2016;**6**(6):a025197

[5] Wright GD. The antibiotic resistome: The nexus of chemical and genetic diversity. Nature Reviews Microbiology. 2007;**5**(3):175

[6] Anderson HW, Gottlieb D. Plant disease control with antibiotics. Economic Botany. 1952;**6**(3):294-308

[7] Leben C, Keitt GW. Antibiotics and plant disease, effects of antibiotics in control of plant diseases. Journal of Agricultural and Food Chemistry. 1954;**2**(5):234-239

[8] McManus PS, Stockwell VO, Sundin GW, Jones AL. Antibiotic use in plant agriculture. Annual Review of Phytopathology. 2002;**40**(1):443-465

[9] McManus PS, Jones AL. Epidemiology and genetic analysis of streptomycin-resistant *Erwinia amylovora* from Michigan and evaluation of oxytetracycline for control. Phytopathology (USA). 1994

[10] Moller WJ, Schroth MN, Thomson SV. The scenario of fire blight and streptomycin resistance [*Erwinia amylovora*; California; USA]. Plant Diseases (USA). 1981

[11] Hu J, Jiang J, Wang N. Control of citrus Huanglongbing (HLB) via trunk injection of plant activators and antibiotics. Phytopathology. 2018;**108**:186-195

[12] Hu J, Wang N. Evaluation of the spatiotemporal dynamics of oxytetracycline and its control effect against citrus Huanglongbing via trunk injection. Phytopathology. 2016;**106**:1495-1503

[13] Wang N, Pierson EA, Setubal JC, Xu J, Levy JG. The *Candidatus liberibacter*–host interface: Insights into pathogenesis mechanisms and disease control. Annual Review of Phytopathology. 2017;**55**:451-482

[14] Shtienberg D, Zilberstaine M, Oppenheim D, Herzog Z, Manulis S. Efficacy of oxolinic acid and other bactericides in suppression of Erwinia amylovora in pear orchards in Israel. Phytoparasitica. 2001;**29**:143-154

[15] Vidaver AM. Use of antimicrobials in plant agriculture. Clinical Infectious Diseases. 2002;**34**(Suppl):S107-S110

[16] Ishiyama T, Hara I, Matsuoka M, Sato K, Shimada S. Studies on preventive effect of kasugamycin on rice blast. The Journal of Antibiotics. 1965;**18**:115-119

[17] McGhee GC, Sundin GW. Evaluation of kasugamycin for fire blight management, effect on non-target bacteria, and assessment of kasugamycin resistance potential in *Erwinia amylovora*. Phytopathology. 2011;**101**:192-204

[18] Chiou C-S, Jones AL. Nucleotide sequence analysis of a transposon (Tn*5393*) carrying streptomycin resistance genes in *Erwinia amylovora* and other gram-negative bacteria. Journal of Bacteriology. 1993;**175**:732-740

[19] Sundin GW, Bender CL. Expression of the *strA-strB* streptomycin resistance genes in *Pseudomonas syringae* and *Xanthomonas campestris* and characterization of IS*6100* in *X. campestris*. Applied and Environmental Microbiology. 1995;**61**:2891-2897

[20] Forster H, McGhee GC, Sundin GW, Adaskaveg JE. Characterization of streptomycin resistance in isolates of *Erwinia amylovora* in California. Phytopathology. 2015;**105**:1302-1310

[21] Palmer EL, Teviotdale BL, Jones AL. Arelative of the broad-host-range plasmid RSF1010 detected in *Erwinia amylovora*. Applied and Environmental Microbiology. 1997;**63**:4604-4607

[22] Ohshima K, Taniyama T, Yamanaka T, Ishikawa M, Naito S. Isolation of a mutant of *Arabidopsis thaliana* carrying two simultaneous mutations affecting tobacco mosaic virus multiplication within a single cell. Virology. 1998;**243**:472-481

[23] Xu Y, Luo Q, Zhou M. Identification and characterization of integron-mediated antibiotic resistance in the phytopathogen *Xanthomonas oryzae* pv. *oryzae*. PLoS ONE. 2013;**8**:e55962

[24] Han HS, Koh YJ, Hur J-S, Jung JS. Occurrence of the *strA-strB* streptomycin resistance genes in *Pseudomonas* species isolated from kiwifruit plants. Journal of Microbiology. 2004;**42**:365-368

[25] McGhee GC, Guasco J, Bellomo LM, Blumer-Schuette SE, Shane WW. Genetic analysis of streptomycin-resistant (SmR) strains of *Erwinia amylovora* suggests that dissemination of two genotypes is responsible for the current distribution of SmR *E. amylovora* in Michigan. Phytopathology. 2011;**192**:182-191

[26] Sundin GW. Examination of base pair variants of the *strA-strB* streptomycin resistance genes from bacterial pathogens of humans, animals, and plants. The Journal of Antimicrobial Chemotherapy. 2000;**46**:848-849

[27] Sundin GW. Distinct recent lineages of the *strA-strB* streptomycin resistance genes in clinical and environmental bacteria. Current Microbiology. 2002;**45**:63-69

[28] Sundin GW, Bender CL. Ecological and genetic analysis of copper and streptomycin resistance in *Pseudomonas syringae* pv. *syringae*. Applied and Environmental Microbiology. 1993;**59**:1018-1024

[29] Sundin GW, Bender CL. Molecular analysis of closely related copper- and streptomycin-resistance plasmids in *Pseudomonas syringae* pv. *syringae*. Plasmid. 1996;**35**:98-107

[30] Tancos KA, Villani S, Kuehne S, Borejsza-Wysocka E, Breth D. Prevalence of streptomycinresistant *Erwinia amylovora* in New York apple orchards. Plant Disease. 2016;**100**:802-809

[31] Rezzonico F, Stockwell VO, Duffy F. Plant agricultural streptomycin formulations do not carry antibiotic resistance genes. Antimicrobial Agents and Chemotherapy. 2009;**53**:3173-3177

[32] Chiou C-S, Jones AL. Molecular analysis of high-level streptomycin resistance in *Erwinia amylovora*. Phytopathology. 1995;**85**:324-328

[33] Schroth MN, Thomson SV, Moller WJ. Streptomycin resistance in

Erwinia amylovora. Phytopathology. 1979;**69**:565-568

[34] Hwang MS, Morgan RI, Sarkar SF, Wang PW, Guttman DS. Phylogenetic characterization of virulence and resistance phenotypes of *Pseudomonas syringae*. Applied and Environmental Microbiology. 2005;**71**:5182-5191

[35] Spotts RA, Cervantes LA. Copper, oxytetracycline, and streptomycin resistance of *Pseudomonas syringae* pv. *syringae* strains from pear orchards in Oregon and Washington. Plant Disease. 1995;**79**:1132-1135

[36] Luo Z-Q, Farrand SK. Cloning and characterization of a tetracycline resistance determinant present in *Agrobacterium tumefaciens* C58. Journal of Bacteriology. 1999;**181**:618-626

[37] Loper JE, Henkels MD, Roberts RG, Grove GG, Willet MJ, Smith TJ. Evaluation of streptomycin, oxytetracycline, and copper resistance of *Erwinia amylovora* isolated from pear orchards in Washington state. Plant Disease. 1991;**75**:287-290

[38] Manulis S, Kleitman F, Dror O, Shabi E. Isolation of strains of *Erwinia amylovora* resistant to oxolinic acid. IOBC WPRS Bulletin. 2000;**23**:89-92

[39] Manulis S, Kleitman F, Shtienberg D, Schwartz H, Oppenheim D. Changes in the sensitivity of *Erwinia amylovora* populations to streptomycin and oxolinic acid in Israel. Plant Disease. 2003;**87**:650-654

[40] Kleitman F, Shtienberg D, Blachinsky D, Oppenheim D, Zilberstaine M. *Erwinia amylovora* populations resistant to oxolinic acid in Israel: Prevalence, persistence and fitness. Plant Pathology. 2005;**54**:108-115

[41] Hikitchi Y, Egami H, Ogure Y, Okino T. Fitness for survival of *Burkholderia glumae* resistant to oxolinic acid in rice plant. Annals of the Phytopathological Society of Japan. 1998;**64**:147-152

[42] Spadafora VJ, Orr G, Wade L, Wiglesworth M. Kasugamycin: A novel antibiotic for North American agriculture. Phytopathology. 2010;**100**:S166

[43] Hori T, Kuroda T, Ishikawa K. Occurrence of kasugamycin-resistant *Burkholderia glumae*. Annals of the Phytopathological Society of Japan. 2007;**73**:278

[44] Takeuchi T, Tamura O. Occurrence of kasugamycin-resistant *Acidovorax avenae* ssp. *avenae*. Annals of the Phytopathological Society of Japan. 1991;**57**:117-118

[45] Yoshii A, Moriyama H, Fukuhara T. The novel kasugamycin 2-*N*-acetyltransferase gene *aac(2)IIa*, carried by the IncP island, convers kasugamycin resistance to rice-pathogenic bacteria. Applied and Environmental Microbiology. 2012;**78**:5555-5564

[46] Yoshii A, Omatsu T, Katayama Y, Koyama S, Mizutani T. Two types of genetic carrier, the IncP genomic island and the novel IncP-1ß plasmid, for the *aac(28.20 Sundin)-IIa* gene that confers kasugamycin resistance in *Acidovorax avenae* ssp. *avenae*. Molecular Plant Pathology. 2015;**16**:288-300

[47] Steiner PW. The distribution of spray materials between target and non-target areas of a mature apple orchard by airblast equipment [MS thesis]. Ithaca, NY: Cornell University; 1969

[48] Christiano RSC, Reilly CC, Miller WP, Scherm H. Oxytetracycline dynamics on peach leaves in relation to temperature, sunlight, and simulated rain. Plant Disease. 2010;**94**:1213-1218

[49] Scherer A, Vogt H-R, Vilei EM, Frey J, Perreten V. Enhanced antibiotic multi-resistance in nasal and fecal bacteria after agricultural use of streptomycin. Environmental Microbiology. 2013;**15**:297-304

[50] Ye Z-L, Deng Y, Lou Y, Ye X, Zhang J, Chen S. Adsorption behavior of tetracyclines by struvite particles in the process of phosphorus recovery from synthetic swine wastewater. Chemical Engineering Journal. 2017;**313**:1633-1638

[51] Tancos KA, Cox KD. Effects of consecutive streptomycin and kasugamycin applications on epiphytic bacteria in the apple phyllosphere. Plant Disease. 2017;**101**:158-164

[52] Shade A, Klimowicz AK, Spear RN, Linske M, Donato JJ, et al. Streptomycin application has no detectable effect on bacterial community structure in apple orchard soil. Applied and Environmental Microbiology. 2013;**79**:6617-6625

[53] Walsh F, Smith DP, Owens SM, Duffy B, Frey JE. Restricted streptomycin use in apple orchards did not adversely affect the soil bacteria communities. Frontiers in Microbiology. 2014;**4**:383

[54] Chen QL, An XL, Zhu YG, Su JQ, Gillings MR. Application of struvite alters the antibiotic resistome in soil, rhizosphere, and phyllosphere. Environmental Science & Technology. 2017;**51**:8149-8157

[55] Barza M, Gorbach SL. The need to improve antimicrobial use in agriculture: Ecological and human health consequences. Clinical Infectious Diseases. 2002;**34**:S71-S144

[56] Thanner S, Drissner D, Walsh F. Antimicrobial resistance in agriculture. MBio. 2016;7:e02227-e02215

[57] Stukenbrock EH, Bataillon T. A population genomics perspective on the emergence and adaptation of new plant pathogens in agro-ecosystem. PLoS Pathogens. 2012;**8**:e1002893

[58] Misra BB, Chaturvedi R. When plants braces for the emerging pathogens. Physiological and Molecular Plant Pathology. 2015;**92**:181-185

[59] Britton KO, Liebhold AM. One world, many pathogens. The New Phytologist. 2013;**197**:9-10

[60] Cobb RC, Filipe JAN, Meentemeyer RK, Gilligan CA, Rizzo DM. Ecosystem transformation by emerging infectious disease: Loss of large tanoak from California forests. Journal of Ecology. 2012;**100**:712-722

[61] Fisher MC, Henk DA, Briggs CJ, Brownstein JS, Madoff LC, McCraw SL. Emerging fungal threats to animal, plant and ecosystem health. Nature. 2012;**484**:186-194

[62] Goss EM. Genome-enabled analysis of plant pathogen migration. Annual Review of Phytopathology. 2015;**53**:121-135

[63] Benson DA, Karsch-Mizrachi I, Lipman DJ, Ostell J, Wheeler DL. GenBank. Nucleic Acids Research. 2006;**34**:D16-D20

[64] Benson DA, Karsch-Mizrachi I, Clark K, Lipman DJ, Ostell J, Sayers EW. GenBank. Nucleic Acids Research. 2012;**40**:D48-D53

[65] Thynne E, McDonald MC, Solomon PS. Phytopathogen emergence in the genomics era. Trends in Plant Science. 2015;**20**:246-255

[66] Stukenbrock EH, Banke S, Javan-Nikkhah M, McDonald BA. Origin and domestication of the fungal wheat pathogen *Mycosphaerella graminicola* via sympatric speciation.

Molecular Biology and Evolution. 2007;**24**:398-411

[67] Keeling PJ, Palmer JD. Horizontal gene transfer in eukaryotic evolution. Nature Reviews. Genetics. 2008;**9**:605-618

[68] Giraud T, Refrégier G, Le Gac M, de Vienne DM, Hood ME. Speciation in fungi. Fungal Genetics and Biology. 2008;**45**:791-802

[69] Raffaele S, Win J, Cano LM, Kamoun S. Analyses of genome architecture and gene expression reveal novel candidate virulence factors in the secretome of *Phytophthora infestans*. BMC Genomics. 2010;**11**:637

[70] Cochrane G, Karsch-Mizrachi I, Nakamura Y. The international nucleotide sequence database collaboration. Nucleic Acids Research. 2011;**39**:D15-D18

[71] Chisholm ST, Coaker G, Day B, Staskawicz BJ. Host–microbe interactions: Shaping the evolution of the plant immune response. Cell. 2006;**124**:803-814

[72] Jones JD, Dangl JL. The plant immune system. Nature. 2006;**444**:323-329

[73] Jones JD, Witek K, Verweij W, Jupe F, Cooke D, Dorling S, et al. Elevating crop disease resistance with cloned genes. Philosophical Transactions of the Royal Society of London. Series B, Biological Sciences. 2014;**369**:20130087

[74] Lacombe S, Rougon-Cardoso A, Sherwood E, Peeters N, Dahlbeck D, van Esse HP, et al. Interfamily transfer of a plant pattern-recognition receptor confers broad-spectrum bacterial resistance. Nature Biotechnology. 2010;**28**:365-369

[75] Tripathi JN, Lorenzen J, Bahar O, Ronald P, Tripathi L. Transgenic expression of the rice Xa21 pattern-recognition receptor in banana (*Musa* sp.) confers resistance to *Xanthomonas campestris* pv. Musacearum. Plant Biotechnology Journal. 2014;**12**:663-673

[76] Schwessinger B, Bahar O, Thomas N, Holton N, Nekrasov V, Ruan D, et al. Transgenic expression of the dicotyledonous pattern recognition receptor EFR in rice leads to ligand-dependent activation of defense responses. PLoS Pathogens. 2015;**11**:e1004809

[77] Dangl JL, Horvath DM, Staskawicz BJ. Pivoting the plant immune system from dissection to deployment. Science. 2013;**341**:746-751

[78] Kim HJ, Chiang YH, Kieber JJ, Schaller GE. SCF(KMD) controls cytokinin signaling by regulating the degradation of type-B response regulators. Proceedings of the National Academy of Sciences of the United States of America. 2013;**110**:10028-10033

[79] Curtis RH, Pankaj, Powers SJ, Napier J, Matthes MC. The Arabidopsis F-box/Kelch-repeat protein At2g44130 is upregulated in giant cells and promotes nematode susceptibility. Molecular Plant-Microbe Interactions. 2013;**26**:36-43

[80] Kessler SA, Shimosato-Asano H, Keinath NF, Wuest SE, Ingram G. Conserved molecular components for pollen tube reception and fungal invasion. Science. 2010;**330**:968-971

[81] Wang Y, Liu C, Li K, Sun F, Hu H. Arabidopsis EIN2 modulates stress response through abscisic acid response pathway. Plant Molecular Biology. 2007;**64**:633-644

[82] Wubben MJ 2nd, Su H, Rodermel SR, Baum TJ. Susceptibility to the sugar beet cyst nematode

is modulated by ethylene signal transduction in *Arabidopsis thaliana*. Molecular Plant-Microbe Interactions. 2001;**14**:1206-1212

[83] Denancé N, Ranocha P, Oria N, Barlet X, Rivière MP, Yadeta KA, et al. Arabidopsis wat1 (walls are thin1)-mediated resistance to the bacterial vascular pathogen, *Ralstonia solanacearum*, is accompanied by cross-regulation of salicylic acid and tryptophan metabolism. The Plant Journal. 2013;*73*(2):225-239

[84] Abuqamar S, Ajeb S, Sham A, Enan MR, Iratni R. A mutation in the expansin-like A2 gene enhances resistance to necrotrophic fungi and hypersensitivity to abiotic stress in *Arabidopsis thaliana*. Molecular Plant Pathology. 2013;**14**:813-827

[85] Lumbreras V, Vilela B, Irar S, Sole M, Capellades M. MAPK phosphatase MKP2 mediates disease responses in Arabidopsis and functionally interacts with MPK3 and MPK6. The Plant Journal. 2010;**63**:1017-1030

[86] Ma X, Browse J. Altered rates of protein transport in Arabidopsis mutants deficient in chloroplast membrane unsaturation. Phytochemistry. 2006;**67**:1629-1636

[87] Levings CS 3rd. The Texas cytoplasm of maize: Cytoplasmic male sterility and disease susceptibility. Science. 1990;**250**:942-947

[88] Hansjakob A, Riederer M, Hildebrandt U. Wax matters: Absence of very-long-chain aldehydes from the leaf cuticular wax of the glossy11 mutant of maize compromises the prepenetration processes of *Blumeria graminis*. Plant Pathology. 2011;**60**:1151-1161

[89] Cantu D, Vicente AR, Greve LC, Dewey FM, Bennett AB. The intersection between cell wall disassembly, ripening, and fruit susceptibility to *Botrytis cinerea*. Proceedings of the National Academy of Sciences of the United States of America. 2008;**105**:859-864

[90] Harrison E, Burbidge A, Okyere JP, Thompson AJ, Taylor IB. Identification of the tomato ABA-deficient mutant sitiens as a member of the ABA-aldehyde oxidase gene family using genetic and genomic analysis. Plant Growth Regulation. 2011;**64**:301-309

[91] Humphry M, Reinstadler A, Ivanov S, Bisseling T, Panstruga R. Durable broad-spectrum powdery mildew resistance in pea er1 plants is conferred by natural loss-of-function mutations in PsMLO1. Molecular Plant Pathology. 2011;**12**:866-878

[92] Sanchez-Hernandez C, Lopez MG, Delano-Frier JP. Reduced levels of volatile emissions in jasmonate-deficient spr2 tomato mutants favor oviposition by insect herbivores. Plant, Cell & Environment. 2006;**29**:546-557

[93] Avila CA, Arevalo-Soliz LM, Jia L, Navarre DA, Chen Z. Loss of function of FATTY ACID DESATURASE7 in tomato enhances basal aphid resistance in a salicylate-dependent manner. Plant Physiology. 2012;**158**:2028-2041

[94] Krasikov V, Dekker HL, Rep M, Takken FL. The tomato xylem sap protein XSP10 is required for full susceptibility to Fusarium wilt disease. Journal of Experimental Botany. 2011;**62**:963-973

[95] Shen X, Liu H, Yuan B, Li X, Xu C, Wang S. OsEDR1 negatively regulates rice bacterial resistance via activation of ethylene biosynthesis. Plant, Cell & Environment. 2011;**34**:179-191

[96] Xiong L, Yang Y. Disease resistance and abiotic stress tolerance in rice are inversely modulated by

an abscisic acid-inducible mitogen-activated protein kinase. Plant Cell. 2003;**15**:745-759

[97] Chujo T, Miyamoto K, Shimogawa T, Shimizu T, Otake Y. OsWRKY28, a PAMP-responsive transrepressor, negatively regulates innate immune responses in rice against rice blast fungus. Plant Molecular Biology. 2013;**82**:23-37

[98] Delteil A, Blein M, Faivre-Rampant O, Guellim A, Estevan J. Building a mutant resource for the study of disease resistance in rice reveals the pivotal role of several genes involved in defense. Molecular Plant Pathology. 2012;**13**:72-82

[99] Jiang CJ, Shimono M, Maeda S, Inoue H, Mori M. Suppression of the rice fatty-acid desaturase gene OsSSI2 enhances resistance to blast and leaf blight diseases in rice. Molecular Plant-Microbe Interactions. 2009;**22**:820-829

[100] Yoshii M, Shimizu T, Yamazaki M, Higashi T, Miyao A. Disruption of a novel gene for a NAC-domain protein in rice confers resistance to rice dwarf virus. The Plant Journal. 2009;**57**:615-625

[101] Yoshii M, Yamazaki M, Rakwal R, Kishi-Kaboshi M, Miyao A, Hirochika H. The NAC transcription factor RIM1 of rice is a new regulator of jasmonate signaling. The Plant Journal. 2010;**61**:804-815

[102] Varallyay E, Giczey G, Burgyan J. Virus-induced gene silencing of Mlo genes induces powdery mildew resistance in *Triticum aestivum*. Archives of Virology. 2012;**157**:1345-1350

[103] Gill US, Lee S, Mysore KS. Host versus nonhost resistance: Distinct wars with similar arsenals. Phytopathology. 2015;**105**:580-587

[104] Holme IB, Wendt T, Holm PB. Intragenesis and cisgenesis as alternatives to transgenic crop development. Plant Biotechnology Journal. 2013;**11**:395-407

[105] Horvath DM, Stall RE, Jones JB, Pauly MH, Vallad GE, Dahlbeck D, et al. Transgenic resistance confers effective field level control of bacterial spot disease in tomato. PLoS One. 2012;**7**:e42036

[106] Tai TH, Dahlbeck D, Clark ET, Gajiwala P, Pasion R, Whalen MC, et al. Expression of the Bs2 pepper gene confers resistance to bacterial spot disease in tomato. Proceedings of the National Academy of Sciences of the United States of America. 1999;**96**:14153-14158

[107] Kawashima CG, Guimarães GA, Nogueira SR, MacLean D, Cook DR, Steuernagel B, et al. A pigeonpea gene confers resistance to Asian soybean rust in soybean. Nature Biotechnology. 2016;**34**(6):661

[108] Kim SH, Qi D, Ashfield T, Helm M, Innes RW. Using decoys to expand the recognition specificity of a plant disease resistance protein. Science. 2016;**351**:684-687

[109] Pel MA, Foster SJ, Park TH, Rietman H, van Arkel G, Jones JD, et al. Mapping and cloning of late blight resistance genes from *Solanum venturii* using an interspecific candidate gene approach. Molecular Plant-Microbe Interactions. 2009;**22**:601-615

[110] Fukuoka S, Saka N, Mizukami Y, Koga H, Yamanouchi U, Yoshioka Y, et al. Gene pyramiding enhances durable blast disease resistance in rice. Scientific Reports. 2015;**5**:7773

[111] Dutt M, Barthe G, Irey M, Grosser J. Transgenic citrus expressing an Arabidopsis NPR1 gene exhibit enhanced resistance against huanglongbing (HLB; citrus greening). PLoS One. 2015;**10**:e0137134

[112] Chen XJ, Chen Y, Zhang LN, Xu B, Zhang JH, Chen ZX, et al. Overexpression of OsPGIP1 enhances rice resistance to sheath blight. Plant Disease. 2016;**100**:388-395

[113] van Schie CC, Takken FL. Susceptibility genes 101: How to be a good host. Annual Review of Phytopathology. 2014;**52**:551-581

[114] Carthew RW, Sontheimer EJ. Origins and mechanisms of miRNAs and siRNAs. Cell. 2009;**136**:642-655

[115] Gonsalves D, Ferreira S. Transgenic papaya: A case for managing risks of papaya ring spot virus in Hawaii. Plant Health Progress. 2003

[116] Noon JB, Hewezi T, Maier TR, Simmons C, Wei JZ, Wu G, et al. Eighteen new candidate effectors of the phytonematode *Heterodera glycines* produced specifically in the secretory esophageal gland cells during parasitism. Phytopathology. 2015;**105**:1362-1372

[117] Baum JA, Bogaert T, Clinton W, Heck GR, Feldmann P, Ilagan O, et al. Control of coleopteran insect pests through RNA interference. Nature Biotechnology. 2007;**25**:1322-1326

[118] Ali Z, Abulfaraj A, Idris A, Ali S, Tashkandi M, Mahfouz MM. CRISPR/Cas9-mediated viral interference in plants. Genome Biology. 2015;**16**:238

[119] Ji X, Zhang H, Zhang Y, Wang Y, Gao C. Establishing a CRISPR–Cas-like immune system conferring DNA virus resistance in plants. Nature Plants. 2015;**1**:15144

[120] Chaparro-Garcia A, Kamoun S, Nekrasov V. Boosting plant immunity with CRISPR/Cas. Genome Biology. 2015;**16**:254

[121] Baltes NJ, Hummel AW, Konecna E, Cegan R, Bruns AN, Bisaro DM, et al. Conferring resistance to geminiviruses with the CRISPR–Cas prokaryotic immune system. Nature Plants. 2015;**1**:15145

[122] Belhaj K, Chaparro-Garcia A, Kamoun S, Nekrasov V. Plant genome editing made easy: Targeted mutagenesis in model and crop plants using the CRISPR/Cas system. Plant Methods. 2013;**9**:39

[123] Doudna JA, Charpentier E. The new frontier of genome engineering with CRISPR-Cas9. Science. 2014;**346**:1258096

[124] Woo JW, Kim J, Kwon SI, Corvalan C, Cho SW, Kim H, et al. DNA-free genome editing in plants with preassembled CRISPR-Cas9 ribonucleoproteins. Nature Biotechnology. 2015;**33**:1162-1164

[125] Hallerman E, Grabau E. Crop biotechnology: A pivotal moment for global acceptance. Food and Energy Security. 2016;**5**:3-17

[126] Voytas DF, Gao C. Precision genome engineering and agriculture: Opportunities and regulatory challenges. PLoS Biology. 2014;**12**:e1001877

[127] Gaj T, Gersbach CA, Barbas CF III. ZFN, TALEN, and CRISPR/Cas-based methods for genome engineering. Trends in Biotechnology. 2013;**31**:397-405

[128] Gaspar YM, Nam J, Schultz CJ, Lee LY, Gilson PR. Characterization of the Arabidopsis lysine-rich arabinogalactan-protein AtAGP17 mutant (rat1) that results in a decreased efficiency of agrobacterium transformation. Plant Physiology. 2004;**135**:2162-2171

[129] Zhu Y, Nam J, Humara JM, Mysore KS, Lee LY. Identification of Arabidopsis rat mutants. Plant Physiology. 2003;**132**:494-505

Section 3

One Health Challenges

Chapter 8

Of Animal and Men: The Importance of Animal Environment to Antimicrobial Resistance: A One Health Approach

Miliane Moreira Soares de Souza,
Cláudio Marcos Rocha-de-Souza, Dayanne Araújo de Melo,
Cássia Couto da Motta, Ramon Loureiro Pimenta,
Irene da Silva Coelho and Shana de Mattos de Oliveira Coelho

Abstract

The contribution of the animal environments to the worsening of the global antimicrobial resistance framework is related to the use of antimicrobials in subtherapeutic doses and, for long periods, establishing ideal conditions for the circulation of resistance genes, which can be transmitted to pathogens adapted to the human microbiota. The study of the animal environment as conducive to the acceleration of resistance evolution is an emerging and critical area for understanding the development and dissemination of resistance genes among the circulating bacteria. The connection between people, animals, and the environment allows us to consider antimicrobial resistance in an approach within the "One Health" concept, which provides a global strategy for expanding collaboration and interdisciplinary communication. This chapter will highlight the emergence of colistin resistance, a great challenge in antimicrobial resistance field. Also, it will focus on some agents included in the priority list of superbugs of the World Health Organization (WHO) or correlated species already identified in veterinary medicine, such as the critical superbugs; priority level 1, Carbapenem-resistant *Acinetobacter baumannii*, Carbapenem-resistant *Pseudomonas aeruginosa*, and ESBL-producing Carbapenemic-resistant Enterobacteriaceae; and the high-priority, level 2, methicillin-resistant *Staphylococcus aureus* (MRSA).

Keywords: one health, *Staphylococcus pseudintermedius*, *Acinetobacter baumannii*, *mec*A gene, *mcr* genes, beta-lactamases

1. Introduction

Global antimicrobial resistance indices are the subject of concern once it has been predicted that nearly 10 million annual deaths will be attributable to resistant pathogen infections by 2050 [1, 2]. The World Health Organization (WHO), the US Center for Disease Control and Prevention (CDC), and the European Center for Disease Prevention and Control (ECDC) classified the emergence and the spread of antimicrobial-resistant bacteria as one of the three major threats to public health in the twenty first century [3].

Importantly, the emergence of resistance is a natural evolutionary response to antimicrobial exposure. Over thousands of years, fungi and bacteria in the natural environment have developed complex mechanisms to prevent their destruction by toxic substances originating from the microbial competition, and these substances have made it possible to synthesize most antibiotics. Therefore, soils should be evaluated as potential reservoirs of antimicrobial-resistant bacteria and should be considered in assessing risk factors that contribute to the global spread of antimicrobial resistance. Moreover, the active collaboration of the human being in the propitiation of this emergency is undeniable due to the increased selection pressure, mainly given by the indiscriminate use of these drugs in human and veterinary medicine [4].

Antimicrobials not only kill sensitive and select resistant bacteria but also influence the mechanisms of genetic variation such as mutation, recombination, transposition, and gene exchange. Such phenomena can be observed from the soil to the intestinal microbiota of humans or animals exposed to antimicrobial underdosing, as the population of commensal microorganisms includes species that are naturally resistant to some antimicrobials. This selective pressure and subsequent imbalance due to the death of sensitive microorganisms allow bacteria with intrinsic or newly acquired resistance to survive and proliferate [5].

Despite this general understanding, the multifactorial origin of the current worldwide antimicrobial resistance scenario makes the picture complex and challenging to intervene. Although studies point to the hospital environment as the main reservoir for the resistance genes of bacteria that colonize and infect humans, the community environment indeed contributes to the establishment of a diverse set of resistance genes [3].

In 2012, Bhullar and colleagues [6] found multiresistant bacteria from an isolated cave microbiome over 4 million years ago in New Mexico, and some of the microorganisms were resistant to up to 14 commercial antibiotics. In another study, the ability of bacteria to use antibiotics as their sole carbon source was detected, making them a significant reservoir of antimicrobial resistance genes [7].

In this context, little is known about the contribution of animal production and veterinary hospital care environments in the maintenance of resistance genes and consequent resistance dissemination. The study of the contribution of various animal-related environments in accelerating the evolution of resistance is an emerging and critical area for understanding its development and as a model for the dissemination of resistance genes among the circulating bacteria. The connection between people, animals, and environment allows for the consideration of antimicrobial resistance within the One Health concept.

2. Distinct animal environment and its impact on antimicrobial resistance

2.1 The poultry production environment as a source of emerging colistin resistance

The increase in antibiotic resistance is now a global concern, including in food-producing animals. They can serve as a reservoir of antibiotic-resistant bacteria and antibiotic resistance determinants that may be transferred to humans [8, 9]. The systematic use of antibiotics in food-producing animals has been increasing the selection pressure for antibiotic-resistant bacteria, especially in Enterobacteriales such as *Escherichia coli* [10]. Furthermore, the emergence of carbapenem-resistant bacteria worldwide and the increased use of polymyxins as "last-line" antibiotics to treat human infections may have contributed to the spread of its resistance [11, 12]. Due to its low price, colistin has been carried on for decades in the poultry industry, worsening this scenario. It is usually administered to the entire flock and mostly used for metaphylaxis and growth promotion in different countries [13].

2.1.1 The silent colistin transferable plasmid-mediated resistance dissemination

The chromosomal polymyxin resistance is most associated with the modification of the lipopolysaccharide (LPS) following the addition of 4-amino-4-deoxy-L-arabinose to lipid A. Modifications of Ara4N are regulated by two-component systems: PhoP/PhoQ, PmrA/PmrB, and MgrB regulator. Mutations in genes involved in the production of these systems may result in lower antibiotic fixation [14]. However, in 2015, a Chinese research group reported the emergence of a transferable plasmid-mediated resistance gene (mcr-1) from human, porcine, and poultry samples, shifting colistin resistance from a contained problem to a global issue [15]. After identification of mcr-1, full scientific attention led to the recognition of multiple mcr-1 variants [16–18] and eight additional mcr genes. Subsequently, the mcr-2 plasmid-mediated colistin resistance gene was detected from poultry, porcine, and bovine *E. coli* in Belgium [19]. A third mobile colistin resistance gene, mcr-3, has been reported in *E. coli*, *Aeromonas* spp., and *Salmonella* spp. isolates from human and animal samples in Asia and Europe [20]. The mcr-4 was detected in *Salmonella enterica* serovar Typhimurium and *E. coli* isolates from animal sources in Italy, Spain, and Belgium [21]. The mcr-5 was detected in poultry and poultry meat isolates of *S. enterica* serovar Paratyphi from porcine *E. coli* in Germany [22]. The sixth mobile colistin resistance gene, mcr-6, was detected in *Moraxella* sp. from porcine in the United Kingdom [23]. The mcr-7 gene was detected in *Klebsiella pneumoniae* in China [9], the mcr-8 gene in *K. pneumoniae* from porcine and human in China [24], and finally, the mcr-9 gene from human in the United States of America [25]. Despite all these reports, a retrospective analysis demonstrated that the mcr-1 gene had been circulating since the 1980s with the earliest isolates from poultry [26]. So, the plasmid-mediated colistin resistance had been around for about 35 years without being detected until 2015. The silent colistin resistance dissemination could partly be explained by the fact that China is by far the leading colistin producer and, at the same time, the largest consumer of its production [15].

Nevertheless, colistin is often added to feed at low doses and used as a growth promoter in different countries. This practice may be the leading cause of the high rate of colistin-resistant bacteria carrying the mcr genes isolated from food-producing animals compared with humans and accelerate the dissemination of mcr genes from animals to humans [15, 27]. Furthermore, the mcr genes may have originated from food-producing animals. The mcr-1 gene was associated with *ISApl1* insertion sequence element, which was first identified in the porcine pathogen *Actinobacillus pleuropneumoniae* [28], and finally, mcr-1-positive strains usually carry *floR* gene conferring resistance to florfenicol, a drug only used in veterinary medicine [10]. The mcr genes have also been found on diverse plasmid backbones (IncI2, IncHI2, IncX4, and pHNSHP45) with high in vitro transfer rates and often harbored together with other resistance determinants, such as β-lactamases [29]. The prevalence data on colistin resistance vary from different countries and continents. Data from two European AMR monitoring from 2014 to 2016 have reported low colistin resistance rates for broilers and chicken meat in Nordic countries [29]. However, studies have shown moderate prevalence in turkey flocks, chicken and turkey meat in Germany [30] and Switzerland [31], and a high prevalence was found in Portugal [32]. In Asia, the prevalence of colistin resistance in poultry is higher than Europe. Different studies have been reported a remarkable increase in colistin resistance frequency in *E. coli* from porcine, poultry, and cattle in all geographic areas of China [33, 34].

2.1.2 Data on colistin resistance in Brazil

In 2015, Brazil overtook China as the world's second largest poultry producer. Nowadays, about 150 countries from all continents consume Brazilian broiler meat, according to the Brazilian Ministry of Agriculture Livestock and Farming [35]. It is noticeable that scientific and technological advancements have transformed poultry from rural farming to full-fledged industry in the last few decades. However, despite this significant expansion, the Brazilian poultry industry is still highly dependent on antibiotic prescription. Prevalence data on colistin resistance in poultry and broiler are overall scarce in South America, including Brazil, in particular, data regarding the plasmid-mediated resistance to colistin [36–38]. In 2016, a Brazilian research group developed a retrospective antimicrobial resistance study and screened 4.620 Enterobacteriales strains isolated from human, animal, food, and environmental samples for the presence of the *mcr*-1 gene. Samples were collected from 2000 to 2016. In this study, *mcr*-1 gene was detected in 16 *E. coli* strains from poultry and porcine isolated between 2012 and 2013. This surveillance showed evidence that *mcr*-1-harboring *E. coli* has been circulating in food-producing animals in Brazil since 2012 [36]. In 2017, the same research group detected the presence of *mcr*-1-harboring *E. coli* strains isolated from commercial chicken meat sold in markets in São Paulo, southeastern Brazil. Most *E. coli* strains exhibited an MDR phenotype and carried IncX4 plasmids, previously identified in human and animal isolates [38].

Furthermore, between 2015 and 2016, Pimenta [39] also detected a high prevalence of *mcr*-1-harboring *E. coli* from broilers and free-range layer hens in several poultry farms in Rio de Janeiro, southeastern Brazil. Most *E. coli* strains carried IncI2, FIB, and B/O plasmids. In November 2016, the MAPA banned the use of colistin as a feed additive for animal growth promotion purposes (regulatory instruction no. 45 [http://www.agricultura.gov.br/]), following the international recommendations of the World Health Organization. Despite this government action, in 2019, our group detected a high prevalence of the *mcr*-1 gene as the only resistance gene in *E. coli* strains isolated from broilers in several poultry farms in Rio de Janeiro (unpublished data). Data suggest that poultry is still an important reservoir to colistin resistance gene *mcr*-1. As poultry meat is an inexpensive source of protein, its impact on transferring resistance cannot be neglected. The overuse of antibiotics will promote the unrestricted expansion and circulation of drug-resistant strains among the human-animal environment. Therefore, continuous surveillance must be of great concern, improving prevalence data in both human and veterinary settings.

2.1.3 Colistin resistance genes in soils

Colistin, also known as polymyxin E, is produced by some strains of *Paenibacillus polymyxa*, a bacterium commonly found in soils associated with plant roots [40]. In some places around the world, the use of poultry litter is an ordinary measure to improve the physical, chemical, and biological properties of soils in agricultural production. However, animal manure, such as poultry litter, a mixture of organic materials including feces, feed, and bedding, is a valuable nutrient-rich soil fertilizer also has been considered an important reservoir of antibiotic residues, antibiotic-resistant bacteria, and antibiotic resistance genes [41, 42]. The enhancement of the concentration and diversity of antibiotic resistance determinants in soils treated with this organic fertilizer is of concern, even considering that untreated soil environments harbor a natural source of both antibiotics and antibiotic resistance genes [43–46]. The colistin resistance *mcr*-1 gene was detected in all soil samples from intensive vegetable production that received poultry litter as organic fertilizer but also in native vegetation areas that comprise a legal reserve at a mountain region of Rio de Janeiro, Brazil, confirming the previous statement that even natural soil environments act as a reservoir of resistance determinant [47].

2.2 Animal production environmental impact on genetic markers mutations: a study of *mec*A gene of *Staphylococcus aureus* isolated from dairy system

Methicillin-resistant *Staphylococcus* (MRS) spp. are important human pathogens that are also a concern in veterinary medicine and animal agriculture. *Staphylococcus* species are present in a wide range of animal species, including dogs, cats, rabbits, horses, cattle, pigs, poultry, and exotic species, both in healthy carriers and as a cause of infection [48–50]. Besides the broad host range distribution and pathogenicity, its significant antimicrobial resistance levels are of great concern [51]. The high antimicrobial resistance level to beta-lactams favors treatment failures and its persistence in the environment. Bacterial resistance mechanisms to this antimicrobial class include a low-affinity penicillin-binding protein 2a (PBP2a) determined by the expression of *the mec*A gene [52]. The phenotypic methicillin-resistant expression does not depend only on the *mec*A gene. This expression is under a more complex control and is only beginning to be better understood since it is expressed in a peculiar and heterogeneous way [53]. Because of this phenotypic heterogeneity, detection of the *mec*A gene is considered the gold standard method for the confirmation of methicillin-resistant isolates by the Clinical Laboratory Institute [54, 55]. However, for samples of animal origin, this proposition is not reliable, since variants of the *mec* gene impair this detection [56, 50].

2.2.1 The mecC *homolog*

In 2011, the report of MRSA strains presenting unusual features in bovine milk samples from the United Kingdom led to the discovery of a novel *mec*A gene named *mec*ALGA251 [57]. This gene presented just 70% similarity at the nucleotide level to the classical *mec*A gene and could not be detected by routine PCR assays targeting the latter [57]. Shortly after its description, the *mec*ALGA251 was isolated from human clinical infections in the United Kingdom, Denmark, and Ireland [58]. It was renamed as *mec*C gene and has been reported from 13 European countries and have been isolated from 14 different host species [59]. Recently, Loncaric et al. [60] reported its occurrence in coagulase-negative staphylococci (CoNS) from various wild and domestic animals. The discovery of the *mecC* gene reinforced the idea of the circulation of gene variants in the animal production environment and the consequent emergence of new methicillin-resistant strains [61]. Until now, the detection of *the mec*C gene is a challenge, and even though there are several reports of *the mec*C gene in *Staphylococcus* species from humans and animals, the puzzling question is that they are all restricted to European countries. In Brazil, the presence of *mec*ALGA251 in the bovine isolates tested negative for *mec*A was investigated, but all isolates also tested negative for *the mec*ALGA251 [50].

2.2.2 A universal primer design experiment

Previous studies [62, 63] reported several phenotypic methicillin-resistant *Staphylococcus* spp. isolates not correlated with the presence of the *mec*A gene. Otherwise, Melo et al. [56] reported the discovery of a *mec*A gene variant from bovine samples containing mutations in the annealing region that does not allow detection of the gene with the already described primers. It was detected that the primer F's annealing site based on the human *S. aureus* mecA gene specified by Murakami et al. [64] presented punctual nucleotide differences that possibly impaired the annealing and amplification of *mec*A gene from the bovine strains. A two-set study was conducted to confirm this hypothesis. Firstly, original primers were synthesized based on the nucleotide sequences of the *mec*A gene of *Staphylococcus aureus* (HE681097). Those primers failed in amplifying the whole *mec*A gene segment in bovine strains. Instead, they did it successfully for human and equine *Staphylococcus* strains. Next, a second-step primer set was based on a sequence of *S. sciuri mec*A gene (AY820253) and only yielded *mec*A gene segments for bovine strains. The multiple alignments of *mec*A gene sequences from bovine, human, and equine origins revealed that bovine ones presented punctual but significant differences leading to the observed impairment of *mec*A gene detection in bovine strains. This divergence of *mec*A gene sequences is a specificity of bovine samples, probably due to some selective pressure in the dairy environment [56].

To validate the newly designed primers, a set of 107 strains was tested for the presence of the *mec*A gene and its bovine variant in *Staphylococcus* spp. isolates from dairy farms in Brazil and Turkey. Seventeen isolates tested positive for the *mec*A variant, nine from Turkey, and eight from Brazil [65]. Recently, a universal PCR primer set was developed and validated to ensure adequate detection of the *mec* genes (classical and variant) [50]. A set of 563 *Staphylococcus* spp. of different animal origins, from the United States of America, and 248 isolates from Brazil, was tested, and 220 (39.1%) were confirmed as MRS by amplification using a classical, variant, and universal primers. The classical *mec*A gene was detected in 201 isolates, being 177 *S. aureus*, whereas the variant *mec*A was detected in 14 isolates, being 2 *S. aureus* and 12 CoNS isolates. These results reinforce that the variant *mec*A is widespread in the animal environment. Surprisingly, a single strain of *S. xylosus* isolated from a porcine nasal swab carried both *mec* genes (classical and variant). The developed universal primer set successfully amplified *mec* genes in 205 isolates, even four isolates that did not amplify any classical or variant *mec*A using conventional primers. It presented sensitivity, specificity, positive predictive, and negative predictive values higher than 90%, comparing to the classical *mec* gene detection. Also, it presented a higher discriminatory power once four isolates just amplified *mec* genes using this primer set. This report is of high relevance once the development of tools to improve MRS diagnosis is crucial for its accurate and rapid identification. The dairy environment represents a considerable challenge in the emergence of new variants of beta-lactam resistance genes due to the frequent use of this antimicrobials class to prevent subclinical mastitis.

2.3 Companion animals environmental impact on antimicrobial resistance

Companion animals are part of human societies around the world [65]. In veterinary medicine clinical practice, diseases such as pyodermitis, external otitis, urinary tract, and respiratory infections are the most frequent causes for the implementation of antibiotic therapy in dogs and cats. Wide-spectrum antimicrobials also prescribed in human medicine are commonly used in these treatments, such as aminopenicillins with beta-lactamase inhibitors, cephalosporins, fluoroquinolones, macrolides, aminoglycosides, and potentiated sulfonamides [66]. As a result, the extensive and indiscriminate use of such antimicrobials in companion animals, coupled with their proximity to humans, gives canine and feline species importance as sources of antimicrobial resistance spread [67]. In the last decade, the escalation of infectious conditions in the veterinary clinic of pet animals related to hitherto unknown or low prevalence agents such as *Acinetobacter calcoaceticus-Acinetobacter baumannii* complex (Acb complex). Parallel to this, the advances in molecular biology applied to bacteriological diagnosis allowed the reclassification of pathogens, and to identify the sharing pathways to virulence and resistance genes between closely related species, as occurs with *Staphylococcus pseudintermedius*, a species reclassified from molecular studies, with some significant gene sharing with *Staphylococcus aureus*, the most recognized species of this genus, of significant importance in human medicine.

2.3.1 Acinetobacter calcoaceticus-Acinetobacter baumannii *complex (Acb complex): an emerging challenge in companion animal environment*

The *Acinetobacter calcoaceticus-Acinetobacter baumannii* complex (Acb complex) is formed by highly genetically related Gram-negative bacteria, which makes species identification difficult through routine laboratory phenotypic methods [68]. The Acb complex comprises *Acinetobacter baumannii* and its close relatives, *A. calcoaceticus*, *A. dijkshoorniae*, *A. lactucae*, *A. nosocomialis*, *A. pittii*, and *A. seifertii* [69]. The clinically relevant species include *A. baumannii*, *A. pittii*, and *A. nosocomialis* [70]. Members of this complex have emerged as opportunistic pathogens causing infections in human and animal health facilities [71]. Infections include pneumonia, especially in ventilated patients; urinary tract infections, especially in patients with urinary catheters; and other infections associated with the use of intravascular catheters [72]. Infections caused by Acb complex agents are difficult to treat since these pathogens have intrinsic resistance to different classes of antimicrobials and also have the ability to acquire additional resistance genes [73]. Infections caused by representatives of the Acb complex has become a growing challenge in clinical routine, both human and animal, especially considering the multiresistant character of these pathogens. Further, there are currently few studies in the field of veterinary medicine that report the occurrence of the other species of this complex, besides *A. baumannii*, as well as the resistance profile.

The analyses developed by our research group have identified all three species of clinical relevance of Acb complex, with the prevalence of *A. pittii*, in samples of animal infectious processes, which has also presented multiresistant profiles. The identified multidrug-resistant isolates were mainly involved in urinary tract infections of dogs and cats, which confirm the real challenge in the veterinary clinical routine. These findings reinforce the need for proper investigation of these agents in the veterinary environment for the adoption of appropriate control and treatment measures. Carbapenemic antimicrobials constitute an excellent alternative for the treatment of infections caused by these pathogens. However, carbapenemases production is one of the biggest challenges in the healthcare system [74]. Agents of Acb complex have become resistant to carbapenems through different mechanisms, including the presence of metallo-beta-lactamases (class B) and the presence or overexpression of OXA (class D) carbapenemases, especially bla_{OXA-23}, bla_{OXA-24}, bla_{OXA-58}, and bla_{OXA-51}. Considering the species *A. baumannii*, the gene bla_{OXA-51} codes for the intrinsic carbapenemase [73]. This additional resistance conferred by OXA-type carbapenemases is commonly grouped into resistance islands located in a region that favors insertion or deletion within the bacterial chromosome [75]. The ISAba1 insertion sequence located in the upstream region of the OXA genes, such as *bla*$_{OXA-23}$, results in increased expression of this gene, observed by an increase in the minimal inhibitory concentration for carbapenems [76]. In addition to antimicrobial resistance, another factor that has favored the emergence of infections caused by species of the Acb complex is related to the biofilm formation capacity of these pathogens, which contributes to their survival in environmental conditions, favoring their persistence in hospital devices, and on different abiotic and biotic surfaces [77]. Biofilm-associated infections require higher doses of antibiotics, resulting in antimicrobial resistance, increased death, prolonged hospital stays, considerable economic loss, and loss of protection for patients [78].

2.3.2 Staphylococcus pseudintermedius*: an underestimated risk for animal and men*

Staphylococcus pseudintermedius was first described as *S. intermedius* [79] based on bacterial isolates from pigeons, dogs, minks, and horses. For decades, *S. intermedius* was considered the leading species of staphylococci associated with skin and soft tissue infections in dogs until it was demonstrated that *S. intermedius* was actually a heterogeneous group of bacteria [80]. Devriese et al. [80] described the *S. pseudintermedius* species through DNA hybridization and 16S rRNA gene sequencing. Subsequent studies evaluated the phenotypic and genotypic diversity in *S. intermedius* and differentiated it into four distinct species: *S. intermedius*, *S. pseudintermedius*, *S. delphini*, and *S. cornubiensis* which are together referred to as the *Staphylococcus intermedius* group (SIG) [81–83]. Since then, *S. pseudintermedius* has been recognized as the common cause of skin infections in dogs, and it has been proposed that all canine isolates should be termed *S. pseudintermedius* unless genotypic typing methods reveal otherwise [84]. This coagulase-positive staphylococci (CoPS) is commensal to the skin and mucosa of healthy dogs, including hair follicles, conjunctival sacs, nares, oral cavity, and perianal region [85]. It is an opportunistic pathogen, capable of causing disease when the natural resistance of the host is suppressed or when the skin barrier is changed [81]. Atopic dermatitis, medical or surgical procedures, and immunosuppressive diseases are examples of predisposing factors to infection [81]. This pathogen is the leading cause of skin and ear infections but may also cause infection in other tissues and cavities and may be transmitted in the community or hospital setting [48, 81, 85]. Besides being the leading cause of canine pyoderma, *S. pseudintermedius* is also frequently isolated from samples of urinary tract infections and may be a complicating factor in immunomodulatory-responsive lymphocytic-plasmacytic pododermatitis [81].

Although dogs are the natural hosts, *S. pseudintermedius* can colonize and infect other animal species, mainly cats [86, 87]. *Staphylococcus pseudintermedius* and *S. aureus* are the species of CoPS that may be composed of the commensal skin microbiota in cats, but there is no consensus in the veterinary literature as to which is dominant and geographic factors should be considered [88]. In these animals, *S. pseudintermedius* can cause tissue infections, rhinitis, nephritis, pneumonia, urinary tract infections, and septicemia [87, 89, 90]. Since the first report of human infection by *S. pseudintermedius* [86], infections have been reported occasionally and are often directly related to close contact with a dog [81]. Pet owners with *S. pseudintermedius* infections and veterinarians are more likely to be nasally colonized by this agent than other individuals [91–93]. *S. pseudintermedius* infections associated with dog bite wounds and post-mastoidectomy, onycholysis, otitis externa, sinusitis, bacteremia, hospital acquired pneumonia, and brain abscess procedures have been described in humans [94–102].

2.3.2.1 *Is methicillin-resistant* Staphylococcus pseudintermedius *(MRSP) the novel MRSA?*

Two cases of methicillin-resistant *S. pseudintermedius* (MRSP) infection have been described in patients with sinusitis, and in one case, at first, *S. pseudintermedius* was misidentified as MRSA [102, 103], suggesting that there may be an underreporting of cases due to the misidentification of the agent. Twenty-four cases of human infections caused by *S. pseudintermedius* were reported in Canada, most of them associated with skin and soft tissue infections, and in three of them, the strain involved was multidrug-resistant [104]. Human MRSP infections in patients without any contact with dogs suggest that humans may eventually be colonized by MRSP [104, 105] and that human-to-human transmission may occur [81]. These reports highlight the importance of *S. pseudintermedius* as a potential emerging pathogen of zoonotic origin and the need for further studies to understand the transmission to humans and to recognize this epidemiological phenomenon.

The relevance of *S. pseudintermedius* as a pathogen is also related to its antimicrobial resistance potential [48]. The inappropriate prescription and use of the same drugs in humans and animals provide a selection of multidrug-resistant (MDR) isolates and consequently compromise the treatment efficacy [106]. Beta-lactam antibiotics are often the first choice of treatment for *Staphylococcus*-associated infections [107], and methicillin-resistant staphylococci (MRS) are an increasing concern. The emergence of MRSP worldwide has become a major problem for small animal veterinary medicine [108] and the infections caused by this agent, a challenge. This resistance is mainly due to two distinct mechanisms: the production of the beta-lactamase enzyme, encoded by the *bla*Z gene, and the production of the additional low-affinity penicillin-binding protein (PBP2a), which is encoded by the *mec*A gene and regulated by *mec*I and *mec*RI genes. PBP2a determines oxacillin/methicillin resistance due to its reduced affinity for beta-lactams and can carry out transpeptidation reactions when normal PBPs are blocked by the drug, allowing peptidoglycan synthesis and conferring resistance to all antimicrobials of the beta-lactam class [109].

The *mec*A gene is located in a mobile genetic element called staphylococcal cassette chromosome (SCC*mec*) chromosomal cassette, which can be transferred via plasmid, transposons, or mobile genetic elements and integrate into the bacterial genome [110, 111]. The *mec* cassette is made up of two main components: the *mec* complex, composed of the IS43 pathogenicity island, the *mec*A gene, and its *mec*I and *mec*RI regulators, and by the *ccr* complex that encodes the chromosome cassette recombinases, which are responsible for the correct excision and consequent integration of this element into the staphylococcal chromosome [112].

The *mec* cassette may carry other genetic elements such as Tn554, pUB110, and pT181, which encode resistance to other classes of antimicrobials. For example, the erm genes, which are responsible for constitutively expressed or induced cross-resistance to macrolides, lincosamides, and streptogramin B (MLSB), are located in Tn554, present in SCC*mec* types II and III [113, 114]. Horizontal transfer of the *mec*A gene into staphylococci and the genetic elements inserted into the SCC*mec* thus resulted in the worldwide spread of oxacillin/methicillin and MDR clones, making it an additional difficulty to control infections caused by these agents [113]. MDR is often observed in MRSP strains [115–117], which also constitute a reservoir of resistance genes for other staphylococci [108] and represents a major problem as the distribution and prevalence of these organisms in animal clinical specimens are relatively unknown, as well as the presence and circulation of genes such as *mec*A and its potential for propagation in companion animals.

2.3.2.2 Staphylococcus pseudintermedius*: genetic diversity and clonal distribution*

In addition to the challenges of identifying *S. pseudintermedius*, there is a need for standardized typing methods that support an epidemiological investigation and monitoring of MRSP [81]. Different techniques have been employed to characterize and determine the genetic diversity among MRSP strains. Pulsed field gel electrophoresis (PFGE), *spa* gene typing, multilocus sequence typing (MLST), and SCC*mec* cassette typing are commonly used. Despite a considerable number of studies seek to understand the population dynamics of *S. pseudintermedius* worldwide, little is known about the clonal distribution patterns of MRSP strains from Africa and South America [118].

In Brazil, the prevalence of methicillin-resistant *Staphylococcus pseudintermedius* (MRSP) as a cause of infectious diseases in companion animals remains unknown. A recent study, developed by Motta [119] provides an overview of the prevalence and characterization of multidrug-resistant MRSP strains from canine and feline clinical samples in Rio de Janeiro. A significative occurrence of multidrug resistance (MDR) in MRSP strains from Brazilian canine and feline clinical was revealed: all MRSP strains analyzed were resistant to seven different antimicrobial classes: fluoroquinolones, phenicols, macrolides, aminoglycosides, aminoglycosides, lincosamides, and tetracyclines. Among these strains, four closely related *spa* types were detected, with predominance of t02. Two clones were identified by the PFGE technique and four closely related strains (groups III and X). MLST typing revealed the presence of three STs/CCs (ST/CC71, ST265/CC258 and ST282/CC45) never reported previously in MRSP strains derived from canine and feline clinical samples from Brazil, with predominance of the worldwide disseminated ST/CC71-*spa*t02-SCC*mec*II-III strain. Comparative analysis of the typing methods used revealed the importance of combining techniques for a broader understanding of the genetic diversity of MRSP. The report highlights the need for further studies to determine the prevalence and characteristics of MRSP from Brazil, supporting preventive and control measures to overcome the antimicrobial resistance.

2.4 β-Lactamase-producing Gram-negative bacteria in a one health approach

Most Enterobacteria pathogens associated with human enteric illness originate from animals and can be transmitted directly to humans or indirectly through animal origin food, contaminated water, or a common reservoir [120]. Currently, β-lactamase-producing strains have been recovered from urban environments, companion/production animals, and animal source foods, which indicate a possible route of dissemination in different ecosystems.

To better understand these links and to identify control measures to reduce the bacterial resistant infections in humans and animals, a One Health approach is needed [121, 122]. The application of a global concept of cross-linking data will improve the prevention, prediction, and control of zoonotic diseases [123, 124].

Undoubtedly, the mobilization of resistance genes through plasmids, transposons, and integrons is intimately linked with widespread of β-lactamases, facilitating the exchange of genetic elements among various bacteria species that can later colonize different hosts and ecosystems and can be spread by different routes [125].

The detection of ESBLs in bacterial isolates of animal origin, such as *Acinetobacter baumannii*, has raised concern regarding the transmission of ESBL genes between human and animal [126]. Also, *E. coli* strains carrying AmpC-β-lactamases have already been reported in healthy and sick animals and food-producing animals [127, 128]. AmpC-hyperproducing *E. coli* was detected in dairy herds in Brazil in 2019. Since there was no previous report of these AMR bacteria in dairy cattle, it was not possible to compare the mutation positions. Nevertheless, many of the positions observed in *E. coli* from beef cattle, broiler, and meat had already been described for human samples. These findings demonstrate a possible transmission route for these bacteria in the food chain and its dissemination through the environment [129].

ESBL or plasmidial AmpC-β-lactamase producers are also frequently resistant to aminoglycosides and fluoroquinolones. The rate of resistance to these antibiotics among *E. coli* isolates of animal origin has been increasingly reported, and the impact of animal-derived broad-spectrum-β-lactamase-producing Gram-negative bacteria on public health has drawing considerable attention worldwide [127].

2.4.1 *β-Lactamases resistance in Gram-negative bacteria*

The most common mechanism of resistance to beta-lactam antibiotics in Gram-negative bacteria is the production of hydrolytic enzymes of antimicrobial agents, including extended-spectrum beta-lactamases (ESBLs) [130]. Two systems of classifying this array of enzymes are in use: the Bush-Jacoby-Medeiros activity-based system [131] and the Ambler system [132] based on nucleotide and amino acid sequence information [133]. The resistance to beta-lactamase inhibitors characterizes the group I (Ambler class C) beta-lactamases (also known as AmpC enzymes). AmpC is mostly found on chromosomes, and its production is inducible. Group 2 (Ambler Class A) beta-lactamases could easily be transmitted into different bacterial cells once plasmids carry them. This group comprises the largest number of characterized enzymes divided into subgroup 2b hydrolyzing penicillins and cephalosporins and its variation 2be (known as "ESBL"). ESBL present a broad spectrum of various antimicrobials as ceftazidime, cefotaxime, and aztreonam. Clavulanic acid exerts potent inhibition towards them. Group 3 (Ambler Class B) enzymes are metalloenzymes capable of destroying carbapenems. Finally, group 4 beta-lactamases contain those unusual penicillinases not inhibited by clavulanic acid, and four of these enzymes exhibit high rates of hydrolysis with carbenicillin or cloxacillin [134].

The spread of extended-spectrum β-lactamase-producing Gram-negative bacteria has dramatically increased worldwide regarding as one of the most important public health threats. Therefore, their appropriate classification and epidemiological data on the main enzymes disseminated in humans, animals, and the environment are of utmost importance.

2.4.2 *An historical approach*

The first plasmid-mediated beta-lactamase in Gram-negative bacteria was reported in Greece in the 1960s. At the end of the 1970s, most *Escherichia coli* (*E. coli*) and *Klebsiella pneumoniae* (*K. pneumoniae*) strains contained ampicillin hydrolyzing β-lactamases mediated by plasmid (TEM-1, TEM-2, and SHV-1). They could be eliminated using third-generation cephalosporins [135]. The emergence of *K. pneumoniae* strains to harbor a gene encoding β-lactamase that hydrolyzes the extended-spectrum cephalosporins was firstly reported by a study from Germany in 1983. Further, in 1986, *K. pneumoniae* strains resistant to the third-generation cephalosporins were detected in France [136, 137]. This resistance was attributed to a new β-lactamase gene, closely related to TEM-1 and TEM-2, and these newly detected enzymes capable of hydrolyzing extended-spectrum beta-lactam antibiotics were named extended-spectrum β-lactamases (ESBLs) [138]. In 1989, a new ESBL family member not belonging to either the TEM or SHV types was reported:

CTX-M type. Its origin has been confirmed to be completely different from that of TEM or SHV ESBL [139]. Nowadays, more than 600 ESBL has been described, the majority belonging to the CTX-M families and TEM-1/2, SHV-1 β-lactamases mutants [140].

2.4.3 The CTX-M-type β-lactamase resistance dissemination

The CTX-M-type β-lactamases can be further differentiated into at least six sub-lineages or groups, namely, CTX-M-1, CTX-M-2, CTX-M-8, CTX-M-9, CTX-M-25, and KLUC [141]. The impressive worldwide spread of CTX-M-producing Gram-negative bacteria turned them to be considered the primary ESBL producers associated with community-acquired infections. The CTX-M family is described as predominant in South America, as well as in Spain and Eastern Europe [142]. Therefore, according to the increasing number of reports describing these enzymes in Brazil, it appears that CTX-M variants are also prevalent in the country compared to TEM and SHV enzymes, prevalent in North America and Western Europe, respectively [143]. In Brazil, CTX-M has been reported in several states; CTX-M-2, CTX-M-8, and CTX-M-9 subtypes are the most prevalent in human samples. In animal production species such as poultry, swine, cattle, and horses, the prevalent enzymes are CTX-M-2, CTX-M-8, and CTX-M-15 [144]. Unfortunately, there are no nationwide surveillance programs on bacterial resistance and its mechanisms, making it difficult to estimate the proportion of ESBL producers [141].

2.4.4 The AmpC-type β-lactamases

Another enzyme group of the β-lactamases type is AmpC. They are relevant enzymes produced constitutively or induced by chromosomal or plasmidial genes expressed by members of Enterobacterales and other Gram-negative bacteria. This class of β-lactamases belongs to the functional groups 1 and C of the Bush and Ambler's classification, respectively [129]. They are often overlooked because they are not within groups 2b or 2b, as CTX-M, TEM, and SHV. AmpC producers hydrolyze almost all β-lactam antibiotics, including cephalosporins, cephamycins, and penicillins, solely or associated with B -lactamase inhibitors, limiting therapeutic options to treat infections caused by these resistant bacteria.

Of major concern is the hyperproduction of this enzyme in *E. coli*. This phenomenon is caused by spontaneous mutations that produce deregulation of *ampC* and is responsible for resistance to first-, second-, and third-generation cephalosporins and to extended-spectrum beta-lactamase inhibitors [145]. Also, some mutations can induce the appearance of an extended-spectrum AmpC (ESAC) that can hydrolyze fourth-generation cephalosporins and carbapenems. Once carbapenems are the choice therapy for Enterobacteria-producing extended-spectrum beta-lactamase infections, the detection of *ampC* production and its control represent an even big challenge.

2.4.5 The carbapenemases

The carbapenem resistance is related to the production of β-lactamases with versatile hydrolytic capacities. Currently, the most important type of class A carbapenemases are KPC enzymes, whereas VIM, IMP, and (particularly) NDM in class B and OXA-48 (and related) in class D are the more relevant enzymes. Most carbapenemases are plasmid-mediated (with genes frequently located in integrons), favoring its dissemination [146]. Since carbapenemase-producing Gram-negative bacteria generally also contain gene coding for other beta-lactam resistance mechanisms, it is not uncommon for organisms to exhibit complex beta-lactam resistance phenotypes. Besides, these organisms often contain other genes that confer resistance to quinolones, aminoglycosides, tetracyclines, sulfonamides, and other families of antimicrobial agents that cause multidrug resistance (AMR) or even pan-resistance. The emergence of new variants and the prevalence of β-lactamases in isolates of community, environmental, and animal origin has demonstrated the complexity of establishing the origin of resistance.

2.4.6 Challenges in detecting the prevalence of β-lactamases

The incidence of large-scale beta-lactamase-producing organisms' spectrum is difficult to determine. There are significant differences between the detection and interpretation methods used by countries and health institutions throughout the study [147]. Considerable phenotypic confirmatory tests for ESBL (2be and 2b) producers have been described in the literature, and all methods utilize the characteristics of ESBL production inhibition by clavulanic acid.

The Clinical and Laboratory Standard Institute (CLSI) recommended test consists of an initial screening by disk diffusion or by the broth dilution method with ceftazidime, ceftriaxone, cefotaxime, cefpodoxime, and aztreonam followed by a phenotypic confirmatory test with cefotaxime and ceftazidime in the presence and absence of clavulanate [54]. The European Antimicrobial Susceptibility Testing Committee (EUCAST) [148] also recommends these tests, but both documents preconize different disk concentrations, and there are also differences in susceptibility zone sizes for consideration of resistance patterns. These factors lead to difficult interlaboratory standardization and consequently to the correct definition of local, regional, and national epidemiological data.

Specifically, regarding AmpC, the Clinical and Laboratory Standards Institute (CLSI) offers no standard test to detect AmpC producer isolates. There are few antimicrobial agents safely effective against these isolates, and many of them are not available or even not approved for animal use. Although different detection methods are available, the lack of international standardization limits the reporting of AmpC by clinical laboratories, which may underestimate this important mechanism of antimicrobial resistance [149].

3. Conclusions

Different environments related to animal production and clinical care can act as a source of the emergence of resistance genes. Studies developed over two decades show that there are relevant peculiarities that must be considered in the detection and understanding of emerging resistance in animal environments to achieve a systemic and practical approach to control antimicrobial resistance worldwide. This chapter discussed some current challenges, the importance of the poultry production environment in the significant emergence of colistin resistance, the development of a universal primer that made it possible to detect a variant of the *mec*A gene in *Staphylococcus aureus* from the dairy environment, and the emergence of *Acinetobacter baumannii-Acinetobacter calcoaceticus* complex and methicillin-resistant *Staphylococcus pseudintermedius* considering companion animals. Finally, the significant dissemination of the resistance mechanism is determined by the production of different classes of beta-lactamases in Gram-negative bacteria in human and animals environments. These concepts allow considering antimicrobial resistance in a One Health approach, which provides a global strategy for expanding collaboration and interdisciplinary communication.

Acknowledgements

We are grateful to the Foundation for Research Support of the State of Rio de Janeiro (FAPERJ), National Council for Scientific and Technological Development (CNPq), and Coordination for the Improvement of Higher Education Personnel (CAPES) for grants that supported this work. We express our sincere thanks to Dr. Catherine Logue, Dr. Lisa Nolan, and Dr. Nicolle Barbieri from Poultry Disease Research Center, University of Georgia, United States of America, for providing us with technical conditions for the development of part of the reported work. We also thank Dr. Helena Maria Neto Ferreira from the Department of Pharmacy, University of Porto, Portugal, for helping us perform the initial beta-lactamases experiments. Our deep gratitude goes to all the former and present staff of the Veterinary Bacteriology Laboratory at Federal Rural University of Rio de Janeiro.

Author details

Miliane Moreira Soares de Souza[1]*, Cláudio Marcos Rocha-de-Souza[2], Dayanne Araújo de Melo[1], Cássia Couto da Motta[1], Ramon Loureiro Pimenta[3], Irene da Silva Coelho[1] and Shana de Mattos de Oliveira Coelho[1]

1 Microbiology and Immunology Department, Institute of Veterinary, Federal Rural University of Rio de Janeiro, Brazil

2 Research Laboratory in Nosocomial Infection, Institute Osvaldo Cruz Foundation, Rio de Janeiro, Brazil

3 Veterinary School, University of Vassouras, Rio de Janeiro, Brazil

*Address all correspondence to: milianemss@gmail.com

References

[1] World Health Organization. WHO publishes list of bacteria for which new antibiotics are urgently needed [Internet]. 2017. Available from: https://www.who.int

[2] O'Neil J. Trackling drug-resistant infections globally: final report and recommendations. The review on antimicrobial resistance [Internet]. 2016. Available from: https://amr-review.org/sites/default/files/160518

[3] Alós JI. Resistencia bacteriana a los antibióticos: Una crisis global. Antibiotic resistance: A global crisis. Enfermedades Infecciosas y Microbiología Clínica. 2015;**33**:692-699. DOI: 10.1016/j.eimc.2014.10.004

[4] Courvalin P. Why is antibiotic resistance a deadly emerging disease? Clinical Microbiology and Infection. 2016;**22**:405-407. DOI: 10.1016/j.cmi.2016.01.012

[5] Holmes AH, Moore LSP, Sundsfjord A, Steinbakk M, Regmi S, Karkey A. Understanding the mechanisms and drivers of antimicrobial resistance. Antimicrobials: Sustainable Access and Sustainable Effectiveness. 2016;**387**(10014):176-187. DOI: 10.1016/S0140-6736(15)00473-0

[6] Bhullar K, Waglechner N, Pawlowski A, Koteva K, Banks E. Antibiotic resistance is prevalent in an isolated cave microbiome. PLoS One. 2012;7(4):e34953. DOI: 10.1371/journal.pone.0034953

[7] Dantas G, Sommer MO, Oluwasegun RD, Church GM. Bacteria subsisting on antibiotics. Science. 2008;**4**(5872):100-103. DOI: 10.1126/science.1155157

[8] World Health Organization. Antibiotic Resistance [Internet]. 2018. Available from: https://www.who.int/news-room/fact-sheets/detail/antibiotic-resistance

[9] Wang Y, Tian GB, Zhang R, et al. Prevalence, risk factors, outcomes, and molecular epidemiology of mcr-1-positive Enterobacteriaceae in patients and healthy adults from China: An epidemiological and clinical study. The Lancet Infectious Diseases. 2017;**17**(4):390-399. DOI: 10.1016/S1473-3099(16)30527-8

[10] Poirel L, Nordmann P. Emerging plasmid-encoded colistin resistance: The animal world as the culprit? The Journal of Antimicrobial Chemotherapy. 2016;**71**(8):2326-2327. DOI: 10.1093/jac/dkw074

[11] Aires CAM, Conceição-Neto OC, Oliveira TRTE, Dias CF, Montezzi LF, Picão RC, et al. Emergence of plasmid-mediated mcr-1 gene in clinical KPC-2-producing *Klebsiella pneumoniae* ST392 in Brazil. Antimicrobial Agents and Chemotherapy. 2017;**50**:282-284. DOI: 10.1128/AAC.00317-17

[12] Conceição-Neto OC, Aires CAM, Pereira NF, da Silva LHJ, Picão RC, Siqueira BN, et al. Detection of the plasmid-mediated mcr-1 gene in clinical KPC-2-producing *Escherichia coli* isolates in Brazil. International Journal of Antimicrobial Agents. 2017;**50**(2):282-284. DOI: 10.1016/j.ijantimicag.2017.05.003

[13] Economou V, Gousia P. Agriculture and food animals as a source of antimicrobial-resistant bacteria. Infection and Drug Resistance. 2015;**8**:49-61. DOI: 10.2147/IDR.S55778

[14] Poirel L, Jayol A, Nordmann P. Polymyxins: Antibacterial activity, susceptibility testing, and resistance mechanisms encoded by plasmids or chromosomes. Clinical Microbiology Reviews. 2017;**30**:557-596. DOI: 10.1128/CMR.00064-16

[15] Liu YY, Wang Y, Walsh TR, et al. Emergence of plasmid-mediated colistin resistance mechanism MCR-1 in animals and human beings in China: A microbiological and molecular biological study. The Lancet Infectious Diseases. 2016;**16**(2):161-168. DOI: 10.1016/S1473-3099(15)00424-7

[16] Di Pilato V, Arena F, Tascini C, Cannatelli A, Henrici De Angelis L, Fortunato S, et al. Mcr-1.2, a new mcr variant carried on a transferable plasmid from a colistin-resistant KPC carbapenemase producing *Klebsiella pneumoniae* strain of sequence type 512. Antimicrobial Agents and Chemotherapy. 2016;**60**(9):5612-5615. DOI: 10.1128/AAC.01075-16

[17] Yang YQ, Li YX, Song T, Yang YX, Jiang W, Zhang AY, et al. Colistin resistance gene mcr-1 and its variant in *Escherichia coli* isolates from chickens in China. Antimicrobial Agents and Chemotherapy. 2017;**61**(5):e01204-e01216. DOI: 10.1128/AAC.01204-16

[18] Lu X, Hu Y, Luo M, Zhou H, Wang X, Du Y, et al. MCR-1.6, a new MCR variant carried by an IncP plasmid in a colistin-resistant *Salmonella enterica* serovar Typhimurium isolate from a healthy individual. Antimicrobial Agents and Chemotherapy. 2017;**61**(5):e02632-e02616. DOI: 10.1128/AAC.02632-16

[19] Basil XB, Lammens C, Ruhal R, Kumar-Singh S, Butaye P, Goossens H, et al. Identification of a novel plasmid-mediated colistin-resistance gene, mcr-2, in *Escherichia coli*. Euro Surveillance. 2016;**21**(27). DOI: 10.2807/1560-7917

[20] Yin W, Li H, Shen Y, Liu Z, Wang S, Shen Z, et al. Novel plasmid-mediated colistin resistance gene mcr-3 in *Escherichia coli*. MBio. 2017;**8**(4):e01166-e01117. DOI: 10.1128/mBio.00543-17

[21] Carattoli A, Villa L, Feudi C, et al. Novel plasmid-mediated colistin resistance mcr-4 gene in *Salmonella* and *Escherichia coli*, Italy 2013, Spain and Belgium, 2015 to 2016. Euro Surveillance. 2017;**22**(31):30589. DOI: 10.2807/1560-7917.ES.2017.22.31.30589

[22] Hammerl JA, Borowiak M, Schmoger S, Shamoun D, Grobbel M, Malorny B, et al. Mcr-5 and a novel mcr-5.2 variant in *Escherichia coli* isolates from food and food-producing animals, Germany, 2010 to 2017. The Journal of Antimicrobial Chemotherapy. 2018;**73**(5):1433-1435. DOI: 10.1093/jac/dky020

[23] Partridge SR, Di Pilato V, Doi Y, Feldgarden M, Haft DH, Klimke W, et al. Proposal for assignment of allele numbers for mobile colistin resistance (mcr) genes. The Journal of Antimicrobial Chemotherapy. 2018;**73**(10):2625-2630. DOI: 10.1093/jac/dky262

[24] Wang X, Wang Y, Zhou Y, Li J, Yin W, Wang S, et al. Emergence of a novel mobile colistin resistance gene, mcr-8, in NDM-producing *Klebsiella pneumoniae*. Emerging Microbes & Infections. 2018;7:1-9. DOI: 10.1038/s41426-018-0124-z

[25] Carroll LM, Gaballa A, Guldimann C, Sullivan G, Henderson LO, Wiedmann M. Identification of novel mobilized colistin resistance gene *mcr-9* in a multidrug-resistant, colistin-susceptible *Salmonella enterica* serotype Typhimurium isolate. Molecular Biology. 2019;**10**(3):e00853-e00819. DOI: 10.1128/mBio.00853-19

[26] Shen Z, Wang Y, Shen Y, Shen J, Wu C. Early emergence of mcr-1 in Escherichia coli from food-producing animals. The Lancet Infectious Diseases. 2016;**16**:293. DOI: 10.1016/S1473-3099(16)00061-X

[27] Rhouma M, Beaudry F, Letellier A. Resistance to colistin: What is the fate

for this antibiotic in pig production? International Journal of Antimicrobial Agents. 2016;**48**(2):119-126. DOI: 10.1016/j.ijantimicag.2016.04.008

[28] Tegetmeyer HE, Jones SC, Langford PR, et al. ISApl1, a novel insertion element of *Actinobacillus pleuropneumoniae*, prevents ApxIV based serological detection of serotype 7 strain AP76. Veterinary Microbiology. 2008;**128**(3-4):342-353. DOI: 10.1016/j.vetmic.2007.10.025

[29] European Medicines Agency - Antimicrobial Advice AdHoc Expert Group. Updated advice on the use of colistin products in animals within the European Union: development of resistance and possible impact on human and animal health. 2016. Available from: http://www.ema.europa.eu/docs

[30] Irrgang A, Roschanski N, Tenhagen B, Grobbel M, Skladnikiewicz-Ziemer T, Thomas K, et al. Prevalence of mcr-1 in *E. coli* from livestock and food in Germany, 2010-2015. PLOS One. 2016;**11**(7):e0159863. DOI: 10.1371/journal.pone.0159863

[31] Zurfluh K, Buess S, Stephan R, Nüesch-Inderbinen M. Assessment of the occurrence of MCR producing Enterobacteriaceae in Swiss and imported poultry meat. Journal of Food Science & Technology. 2016;**1**:24-31. DOI: 10.15436/JFST.1.4.5

[32] Manageiro V, Clemente L, Graça R, Correia I, Albuquerque T, Ferreira E, et al. New insights into resistance to colistin and third-generation cephalosporins of *Escherichia coli* in poultry, Portugal: Novel blaCTX-M-166 and blaESAC genes. International Journal of Food Microbiology. 2017;**263**:67-73. DOI: 10.1016/j.ijfoodmicro.2017.10.007

[33] Liu BT, Song F, Zou M, Zhang QD, Shan H. High incidence of Escherichia coli strains coharboring mcr-1 and blaNDM from chickens. Antimicrobial Agents and Chemotherapy. 2017;**61**. DOI: 10.1128/AAC.02347-16

[34] Zhang P, Shen Z, Zhang C, Song L, Wang B, Shang J, et al. Surveillance of antimicrobial resistance among Escherichia coli from chicken and swine, China, 2008-2015. Veterinary Microbiology. 2017;**03**:49-55. DOI: 10.1016/j.vetmic.2017.02.008

[35] Ministry of Agriculture, Livestock and Supply [Internet]. Normative Instruction. 2016. Available from: http://www.in.gov.br/materia/asset_publisher/

[36] Fernandes MR, Moura Q, Sartori L, Silva KC, Cunha MPV, Esposito F, et al. Silent dissemination of colistin-resistant *Escherichia coli* in South America could contribute to the global spread of the mcr-1 gene. Euro Surveillance. 2016;**21**(17)

[37] Lentz SAM, de Lima-Morales D, Cuppertino VML, Nunes LS, da Motta AS, Zavascki AP, et al. Letter to the editor: Escherichia coli harbouring mcr-1 gene isolated from poultry not exposed to polymyxins in Brazil. Euro Surveillance. 2016;**21**:26

[38] Monte DF, Mem A, Fernandes MR, Cerdeira L, Esposito F, Galvão GA, et al. Chicken meat as a reservoir of Colistin-resistant Escherichia coli strains carrying mcr-1 genes in South America. Antimicrobial Agents and Chemotherapy. 2017;**61**(5). DOI: 10.1128/AAC.02718-16

[39] Pimenta RL. Evaluation of antimicrobial resistance and virulence in bacterial poultry isolated in cutting and posture establishments in the state of Rio de Janeiro [thesis]. Rio de Janeiro: Federal Rural University of Rio de Janeiro; 2018

[40] Grady EN, MacDonald J, Liu L, Richman A, Yuan ZC. Current

knowledge and perspectives of *Paenibacillus*: A review. Microbial Cell Factories. 2016;**15**:203-221. DOI: 10.1186/s12934-016-0603-7

[41] Faldynova M, Videnska P, Havlickova H, Sisak F, Juricova H, Baba V, et al. Prevalence of antibiotic resistance genes in faecal samples from cattle, pigs and poultry. Veterinární Medicína. 2013;**58**:298-304

[42] Udikovic-Kolic N, Wichmann F, Broderick NA, Handelsman J. Bloom of resident antibiotic-resistant bacteria in soil following manure fertilization. Proceedings of the National Academy of Sciences. 2014;**111**:15202-15207. DOI: 10.1073/pnas.1409836111

[43] Wu N, Qiao M, Zhang B, Cheng WD, Zhu YG. Abundance and diversity of tetracycline resistance genes in soils adjacent to representative swine feedlots in China. Environmental Science & Technology. 2010;**44**:6933-6939. DOI: 10.1021/es1007802

[44] Zhu Y, Johnson T, Su J, Qiao M, Guo G, Stedtfeld R, et al. Diverse and abundant antibiotic resistance genes in Chinese swine farms. Proceedings of the National Academy of Sciences. 2013;**1109**:3435-3440. DOI: 10.1073/pnas.1222743110

[45] Herath EM, Palansooriya AGKN, Dandeniya WS, Jinadasa RN. An assessment of antibiotic-resistant bacteria in poultry litter and agricultural soils in Kandy district, Sri Lanka. Tropical Agricultural Research. 2016;**27**:389-398. DOI: 10.4038/tar.v27i4.8215

[46] Yang Q, Zhang H, Guo Y, Tian T. Influence of chicken manure fertilization on antibiotic-resistant bacteria in soil and the endophytic bacteria of pakchoi. International Journal of Environmental Research and Public Health. 2016;**13**:662. DOI: 10.3390/ijerph13070662

[47] Oliveira CC, Lopes ES, Barbosa DR, Pimenta RL, Sobrinho NMBA, Coelho SMO, et al. Occurrence of the colistin resistance *mcr*-1 gene in soils from intensive vegetable production and native vegetation. European Journal of Soil Science. 2019;**70**(4):876-881. DOI: 10.1111/ejss.12832

[48] Weese JS. Methicillin resistance *Staphylococcus aureus* in animals. ILAR Journal. 2010;**51**:233-244. DOI: 10.1093/ilar.51.3.233

[49] Calazans-Silva AC, Medeiros PT, Araujo DM, et al. Genetic analysis of mecA gene and detection of homologue pbpD in *Staphylococcus sciuri* group. Brazilian Journal of Microbiology. 2014;**45**(2):651-655. DOI: 10.1590/s1517-83822014000200038

[50] Melo DA, Soares BS, Motta CC, Dubenczuck FC, Barbieri NL, Logue CM, et al. Accuracy of PCR universal primer for methicillin-resistant *Staphylococcus* and comparison of different phenotypic screening assays. Brazilian Journal of Microbiology. Mar 2020;**51**(1):403-407. DOI: 10.1007/s42770-019-00171-6

[51] Pantosti A. Methicillin-resistant *Staphylococcus aureus* associated with animals and its relevance to human health. Frontiers in Microbiology. 2012;**3**:127. DOI: 10.3389/fmicb.2012.00127

[52] Fishovitz J, Hermoso JA, Chang M, Mobashery S. Penicillin-binding protein 2a of methicillin-resistant *Staphylococcus aureus*. IUBMB Life. 2014;**66**:572-577. DOI: 10.1002/iub.1289

[53] Aedo S, Tomasz A. Role of the stringent stress response in the antibiotic resistance phenotype of methicillin-resistant *Staphylococcus aureus*. Antimicrobial Agents and Chemotherapy. 2016;**60**(4):2311-2317. DOI: 10.1128/AAC.02697-15

[54] Clinical and Laboratory Standards Institute (CLSI). Performance Standards for Antimicrobial Susceptibility Testing. 28th ed. Wayne, PA: CLSI, CLSI Supplement M100; 2018

[55] Clinical and Laboratory Standards Institute (CLSI). Performance Standards for Antimicrobial Disk and Dilution Susceptibility Tests for Bacteria Isolated from Animals. 5th ed. Wayne, PA: CLSI, CLSI Standard VET01; 2018

[56] Melo DA, Coelho IS, Motta CC, Rojas ACCM, Dubenczuk FC, Coelho SMO, et al. Impairments of mecA gene detection in bovine *Staphylococcus* spp. Brazilian Journal of Microbiology. 2014;**45**:1075-1082. DOI: 10.1590/S1517-83822014000300041

[57] Garcia-Alvarez L, Holden MT, Lindsay H, Webb CR, Brown DF, Curran MD, et al. Methicillin-resistant *Staphylococcus aureus* with a novel *mec*A homologue in human and bovine populations in the UK and Denmark: A descriptive study. The Lancet Infectious Diseases. 2011;**11**:595-603. DOI: 10.1016/S1473-3099(11)70126-8

[58] Shore AC, Deasy EC, Slickers P, Brennan GO, Connell B, Monecke S, et al. Detection of staphylococcal cassette chromosome *mec* type XI carrying highly divergent *mec*A, *mec*I, *mec*R1, *bla*Z, and *ccr* genes in human clinical isolates of clonal complex 130 methicillin-resistant *Staphylococcus aureus*. Antimicrobial Agents and Chemotherapy. 2011;**55**:3765-3773. DOI: 10.1128/AAC.00187-11

[59] Paterson GK, FJE M, Harrison EM, EJP C, Torok ME, Zadoks RN, et al. The emergence of *mec*C methicillin-resistant *Staphylococcus aureus*. Trends in Microbiology. 2014;**22**:42-47. DOI: 10.1016/j.tim.2013.11.003

[60] Loncaric I, Kübber-Heiss A, Posautz A, Ruppitsch W, Lepuschitz S, Schauer B, et al. Characterization of *mec*C gene-carrying coagulase-negative *Staphylococcus* spp. isolated from various animals. Veterinary Microbiology. 2019;**230**:138-144. DOI: 10.1016/j.vetmic.2019.02.014

[61] Fitzgerald JR. Livestock-associated *Staphylococcus aureus*: Origin, evolution and public health threat. Trends in Microbiology. 2010;**20**:192-198. DOI: 10.1016/j.tim.2012.01.006

[62] Soares LC, Pereira IA, Pribul BR, Oliva MS, Coelho SMO, Souza MMS. Antimicrobial resistance and detection of *mec*A and *bla*Z genes in coagulase-negative *Staphylococcus* isolated from bovine mastitis. Pesquisa Veterinaria Brasileira. 2012;**32**:692-696. DOI: 10.1590/S0100-736X2012000800002

[63] Mendonça ECL, Marques VF, Melo DA, Alencar TA, Coelho IS, Coelho SMO, et al. Phenogenotypical characterization of antimicrobial resistance in *Staphylococcus* spp. isolated from bovine mastitis. Pesquisa Veterinaria Brasileira. 2012;**31**:859-864. DOI: 10.1590/S0100-736X2012000900008

[64] Murakami KW, Minamide K, Wada W, Nakamura E, Teraoka H, Watanbe S. Identification of methicillin resistant strains of staphylococci by polymerase chain reaction. Journal of Clinical Microbiology. 1991;**29**:2240-2244

[65] Amiot C, Bastian B, Martens P. People and companion animals: It takes two to tango. Bioscience. 2016;**66**(7):552-560. DOI: 10.1093/biosci/biw051

[66] Guardabassi L, Damborg P, Stamm I, Kopp PA, Broens EM, Toutain PL. Diagnostic microbiology in veterinary dermatology: Present and future. Veterinary Dermatology. 2017;**28**:146-e30. DOI: 10.1111/vde.12414

[67] Tamang MD, Nam HM, Kim TS, Jang GC, Jung SC, Lim SK. Emergence of extended-spectrum beta-lactamase (CTX-M-15 and CTX-M-14)-producing nontyphoid *Salmonella* with reduced susceptibility to ciprofloxacin among food animals and humans in Korea. Journal of Clinical Microbiology. 2011;**49**(7):2671-2675. DOI: 10.1128/JCM.00754-11

[68] Marí-Almirall M, Cosgaya C, Higgins PG, Van Assche A, Telli M, Huys G, et al. MALDI-TOF/MS identification of species from the *Acinetobacter baumannii* (Ab) group revisited: Inclusion of the novel *A. seifertii* and *A. dijkshoorniae* species. Clinical Microbiology and Infection. 2017;**23**:210.e1-210.e9. DOI: 10.1016/j.cmi.2016.11.020

[69] Püntener-Simmen S, Zurfluh K, Schmitt S, Stephan R, Nüesch-Inderbinen M. Phenotypic and genotypic characterization of clinical isolates belonging to the *Acinetobacter calcoaceticus-Acinetobacter baumannii* (ACB) complex isolated from animals treated at a veterinary Hospital in Switzerland. Frontiers in Veterinary Science. 2019;**6**:17. DOI: 10.3389/fvets.2019.00017

[70] Chen Y, Ai L, Guo P, Huang H, Wu Z, Liang X, et al. Molecular characterization of multidrug resistant strains of *Acinetobacter baumannii* isolated from pediatric intensive care unit in a Chinese tertiary hospital. BMC Infectious Diseases. 2018;**18**:614. DOI: 10.1186/s12879-018-3511-0

[71] Kuzi S, Blum SE, Kahane N, Adler A, Hussein O, Segev G, et al. Multi-drug-resistant *Acinetobacter calcoaceticus-Acinetobacter baumannii* complex infection outbreak in dogs and cats in a veterinary hospital. The Journal of Small Animal Practice. 2016;**57**:617-625. DOI: 10.1111/jsap.12555

[72] Maragakis LL, Perl TM. *Acinetobacter baumannii:* Epidemiology, antimicrobial resistance, and treatment options. Clinical Infectious Diseases. 2008;**2008**(46):1254-1263. DOI: 10.1086/529198

[73] Poirel L, Bonnin RA, Nordmann P. Genetic basis of antibiotic resistance in pathogenic *Acinetobacter* species. IUBMB Life. 2011;**63**:1061-1067. DOI: 10.1002/iub.532

[74] Rodríguez CH, Nastro M, Famiglietti A. Carbapenemases in *Acinetobacter baumannii*. Review of their dissemination in Latin America. Revista Argentina de Microbiología. 2018;**50**(3):327-333. DOI: 10.1016/j.ram.2017.10.006

[75] Héritier C, Poirel L, Fournier PE, Claverie JM, Raoult D, Nordmann P. Characterization of the naturally occurring oxacillinase of *Acinetobacter baumannii*. Antimicrobial Agents and Chemotherapy. 2005;**49**:4174-4179. DOI: 10.1128/AAC.49.10.4174-4179.2005

[76] Corvec S, Poirel L, Naas T, Drugeon H, Nordmann P. Genetics and expression of the carbapenem hydrolyzing oxacillinase gene blaOXA-23 in *Acinetobacter baumannii*. Antimicrobial Agents and Chemotherapy. 2007;**51**:1530-1533. DOI: 10.1128/AAC.01132-06

[77] McQueary CN, Actis LA. *Acinetobacter baumannii* biofilms: Variations among strains and correlations with other cell properties. Journal of Microbiology. 2011;**49**:243-250. DOI: 10.1007/s12275-011-0343-7

[78] Sun F, Qu F, Ling Y, et al. Biofilm-associated infections: Antibiotic resistance and novel therapeutic strategies. Future Microbiology. 2013;**8**(7):877-886. DOI: 10.2217/fmb.13.58

[79] Hajek V. *Staphylococcus intermedius*, a new species isolated from animals. International Journal of Systematic Bacteriology. 1976;**26**:401-408. DOI: 10.1099/00207713-26-4-401

[80] Devriese L, Vancanney M, Baele M, Vaneechoutte M, De Graef E, Snauwaert C, et al. *Staphylococcus pseudintermedius* sp. nov., a coagulase-positive species from animals. International Journal of Systematic and Evolutionary Microbiology. 2005;**55**:1569-1573. DOI: 10.1099/ijs.0.63413-0

[81] Bannoehr J, Guardabassi L. *Staphylococcus pseudintermedius* in the dog: Taxonomy, diagnostics, ecology, epidemiology and pathogenicity. Veterinary Dermatology. 2012;**23**:253-e52. DOI: 10.1111/j.1365-3164.2012.01046

[82] Sasaki T, Kikuchi K, Tanaka Y, Takahashi N, Kamata S, Hiramatsu K. Reclassification of phenotypically identified *Staphylococcus intermedius* strains. Journal of Clinical Microbiology. 2007;**45**:2770-2778. DOI: 10.1128/JCM.00360-07

[83] Murray AK, Lee J, Bendal LR, Zhang L, Sunde M, Schau Slettemeås J, et al. *Staphylococcus cornubiensis* sp. nov., a novel member of the *Staphylococcus intermedius* group (SIG). International Journal of Systematic and Evolutionary Microbiology. 2018;**68**:3404-3408. DOI: 10.1099/ijsem.0.002992

[84] Devriese LA, Hermans K, Baele M, Haesebrouck F. *Staphylococcus pseudintermedius* versus *Staphylococcus intermedius*. Veterinary Microbiology. 2009;**133**:206-207. DOI: 10.1016/j.vetmic.2008.06.002

[85] Van Duijkeren E, Catry B, Greko C, Moreno MA, Pomba MC, Pyörälä S, et al. Review on methicillin-resistant *Staphylococcus pseudintermedius*. The Journal of Antimicrobial Chemotherapy. 2011;**66**:2705-2714. DOI: 10.1093/jac/dkr367

[86] Van Hoovels L, Vankeerberghen A, Boel A, Van Vaerenbergh K, De Beenhouwer H. First case of *Staphylococcus pseudintermedius* infection in a human. Journal of Clinical Microbiology. 2006;**44**:4609-4612. DOI: 10.1128/JCM.01308-06

[87] Kadlec K, Schwarz S, Perreten V, Andersson UG, Finn M, Greko C, et al. Molecular analysis of methicillin-resistant *Staphylococcus pseudintermedius* of feline origin from different European countries and North America. The Journal of Antimicrobial Chemotherapy. 2010;**65**(8):1826-1828. DOI: 10.1093/jac/dkq203

[88] Abraham JL, Morris DO, Griffeth GC, Shofer FS, Rankin SC. Surveillance of healthy cats and cats with inflammatory skin disease for colonization of the skin by methicillin-resistant coagulase-positive staphylococci and *Staphylococcus schleiferi* ssp. *schleiferi*. Veterinary Dermatology. 2001;**18**:252-259. DOI: 10.1111/j.1365-3164.2007.00604.x

[89] Wettstein K, Descloux S, Rossan A, Perreten V. Emergence of methicillin-resistant *Staphylococcus pseudintermedius* in Switzerland: Three cases of urinary tract infections in cats. Schweizer Archiv für Tierheilkunde. 2008;**150**:339-343. DOI: 10.1024/0036-7281.150.7.339

[90] Pomba C, Couto N, Moodley A. Treatment of a lower urinary tract infection in a cat caused by a multi-drug methicillin-resistant *Staphylococcus pseudintermedius* and *Enterococcus faecalis*. Journal of Feline Medicine and Surgery. 2010;**12**:802-806. DOI: 10.1016/j.jfms.2010.04.006

[91] Frank LA, Kania SA, Hnilica KA, Wilkes RP, Bemis DA. Isolation of

Staphylococcus schleiferi from dogs with pyoderma. Journal of the American Veterinary Medical Association. 2003;**222**:451-454. DOI: 10.2460/javma.2003.222.451

[92] Ishihara K, Shimokubo N, Sakagami A, Ueno H, Muramatsu Y, Kadosawa T, et al. Occurrence and molecular characteristics of methicillin-resistant *Staphylococcus aureus* and methicillin-resistant *Staphylococcus pseudintermedius* in an academic veterinary hospital. Applied and Environmental Microbiology. 2010;**76**:5165-5174. DOI: 10.1128/AEM.02780-09

[93] Walther B, Hermes J, Cuny C, Wieler LH, Vincze S, Abou Elnaga Y, et al. Sharing more than friendship – Nasal colonization with coagulase-positive staphylococci (CPS) and co-habitation aspects of dogs and their owners. PLoS One. 2012;**7**:e35197. DOI: 10.1371/journal.pone.0035197

[94] Barnham M, Holmes B. Isolation of CDC group M-5 and *Staphylococcus intermedius* from infected dog bites. The Journal of Infection. 1992;**25**:332-334. DOI: 10.1016/0163-4453(92)91759-5

[95] Lee J. *Staphylococcus intermedius* isolated from dog-bite wounds. The Journal of Infection. 1994;**29**:105. DOI: 10.1016/s0163-4453(94)95276-0

[96] Vandenesch F, Celard M, Arpin D, Bes M, Greenland T, Etienne J, et al. Catheter-related bacteremia associated with coagulase-positive *Staphylococcus intermedius*. Journal of Clinical Microbiology. 1995;**33**:2508-2510

[97] Gerstadt K, Daly JS, Mitchell M, Wessolossky M, Cheeseman SH. Methicillin-resistant *Staphylococcus intermedius* pneumonia following coronary artery bypass grafting. Clinical Infectious Diseases. 1999;**29**:218-219. DOI: 10.1086/520168

[98] Tanner MA, Everett CL, Youvan DC. Molecular phylogenetic evidence for noninvasive zoonotic transmission of *Staphylococcus intermedius* from a canine pet to a human. Journal of Clinical Microbiology. 2000;**38**:1628-1631

[99] Kikuchi K, Karasawa T, Piao C, Itoda I, Hidai H, Yamaura H, et al. Molecular confirmation of transmission route of *Staphylococcus intermedius* in mastoid cavity infection from dog saliva. Journal of Infection and Chemotherapy. 2004;**10**:46-48. DOI: 10.1007/s10156-003-0281-3

[100] Pottumarthy S, Schapiro JM, Prentice JL, Houze YB, Swanzy SR, Fang FC, et al. Clinical isolates of *Staphylococcus intermedius* masquerading as methicillin-resistant *Staphylococcus aureus*. Journal of Clinical Microbiology. 2004;**42**:5881-5884. DOI: 10.1128/JCM.42.12.5881-5884.2004

[101] Atalay B, Ergin F, Cekinmez M, Caner H, Altinors N. Brain abscess caused by *Staphylococcus intermedius*. Acta Neurochirurgica. 2005;**147**:347-348. DOI: 10.1007/s00701-004-0437-7

[102] Kempker R, Mangalat D, Kongphet-Tran T, Eaton M. Beware of the pet dog: A case of *Staphylococcus intermedius* infection. The American Journal of the Medical Sciences. 2009;**338**:425-427. DOI: 10.1097/MAJ.0b013e3181b0baa9

[103] Stegmann R, Burnens A, Maranta CA, Perrten V. Human infection associated with methicillin-resistant *Staphylococcus pseudintermedius* ST71. The Journal of Antimicrobial Chemotherapy. 2010;**65**:2047-2048. DOI: 10.1093/jac/dkq241

[104] Somayaji R, Rubin JE, Priyantha MAR, Church D. Exploring *Staphylococcus pseudintermedius:* An emerging zoonotic pathogen? Future

Microbiology. 2016;**11**:11. DOI: 10.2217/fmb-2016-0137

[105] Kelesidis T, Tsiodras S. *Staphylococcus intermedius* is not only a zoonotic pathogen, but may also cause skin abscesses in humans after exposure to saliva. International Journal of Infectious Diseases. 2010;**14**:e838-e841. DOI: 10.1016/j.ijid.2010.02.2249

[106] Ventola CL. The antibiotic resistance crisis part 1: Causes and threats. Pharmacy and Therapeutics. 2015;**40**(4):277-283

[107] Chatterjee SS, Otto M. Improved understanding of factors driving methicillin-resistant *Staphylococcus aureus* epidemic waves. Clinical Epidemiology. 2013;**5**:205-217. DOI: 10.2147/CLEP.S37071

[108] Pomba C, Rantala M, Greko C, Baptiste KE, Catry B, Van Duijkeren E, et al. Public health risk of antimicrobial resistance transfer from companion animals. The Journal of Antimicrobial Chemotherapy. 2017;**72**(4):957-968. DOI: 10.1093/jac/dkw481

[109] Pehlivanoglu F, Yardimci H. Detection of methicillin and vancomycin resistance in *Staphylococcus* strains isolated from bovine milk samples with mastitis. Kafkas Universitesi Veteriner Fakultesi Dergisi. 2012;**18**(5):849-855

[110] Katayama Y, Ito T, Hiramatsu K. Genetic organization of the chromosome region surrounding *mec*A in clinical staphylococcal strains: Role of IS431-mediated *mec*I deletion in expression of resistance in *mec*A-carrying, low-level methicillin-resistant *Staphylococcus haemolyticus*. Antimicrobial Agents and Chemotherapy. 2001;**45**(7):1955-1963. DOI: 10.1128/AAC.45.7.1955-1963.2001

[111] Turlej A, Hryniewicz W, Empel J. Staphylococcal cassette chromosome *mec* (SCC*mec*) classification and typing methods: An overview. Polish Journal of Microbiology. 2011;**60**(2):95-103

[112] Hiramatsu K, Katayama Y, Matsuo M, Sasaki T, Morimoto Y, Sekiguchi A, et al. Multi-drug-resistant *Staphylococcus aureus* and future chemotherapy. The Journal of Infusional Chemotherapy. 2014;**20**(10):593-601. DOI: 10.1016/j.jiac.2014.08.001

[113] Ito T, Katayama Y, Asda K, Mori N, Tsutsumimoto K, Tiensasitorn C, et al. Structural comparison of three types of staphylococcal cassette chromosome *mec* integrated in the chromosome in methicillin-resistant *S. aureus*. Antimicrobial Agents and Chemotherapy. 2001;**45**:1323-1336. DOI: 10.1128/AAC.45.5.1323-1336.2001

[114] Roberts MC. Update on macrolide-lincosamide-streptogramin, ketolide, and oxazolidinone resistance genes. FEMS Microbiology Letters. 2008;**282**:147-159. DOI: 10.1111/j.1574-6968.2008.01145.x

[115] Perreten V, Kadlec K, Schwarz S, GröNlund Andersson U, Finn M, Greko C, et al. Clonal spread of methicillin-resistant *Staphylococcus pseudintermedius* in Europe and North America: An international multicentre study. Journal of Antimicrobial Chemotherapy. 2010;**65**:1145-1154. DOI: 10.1093/jac/dkq078

[116] Ruscher CA, Lubke-Becker A, Semmler T, Wleklinski CG, Paasch A, Soba A, et al. Widespread rapid emergence of a distinct methicillin- and multidrug-resistant *Staphylococcus pseudintermedius* (MRSP) genetic lineage in Europe. Veterinary Microbiology. 2010;**144**:340-346. DOI: 10.1016/j.vetmic.2010.01.008

[117] Moodley A, Stegger M, Zakour NLB, Fitzgerald JR, Guardabassi L. Tandem repeat sequence analysis of staphylococcal protein a

(spa) gene in methicillin-resistant *Staphylococcus pseudintermedius*. Veterinary Microbiology. 2009;**135**:320-3226. DOI: 10.1016/j.vetmic.2008.09.070

[118] Dos Santos TP, Damborg P, Moodley A, Guardabassi L. Systematic review on global epidemiology of methicillin-resistant *Staphylococcus pseudintermedius*: Inference of population structure from multilocus sequence typing data. Frontiers in Microbiology. 2016;**7**:1599. DOI: 10.3389/fmicb.2016.01599

[119] Motta CC. Diversity and antimicrobial resistance in methicillin-resistant Staphylococcus pseudintermedius (MRSP) from infectious processes in companion animals [Thesis]. Rio de Janeiro: Federal Rural University of Rio de Janeiro; 2018

[120] Newell DG, Koopmans M, Verhoef L, Duizer E, Aidara-Kane A, Sprong H, et al. Food-borne diseases – The challenges of 20 years ago still persist while new ones continue to emerge. International Journal of Food Microbiology. 2010;**139**(Suppl 1): S3-S15. DOI: 10.1016/j.ijfoodmicro.2010.01.021

[121] Parmley EJ, Pintar K, Majowicz S, Avery B, Cook A, Jokinen C, et al. A Canadian application of one health: Integration of *Salmonella* data from various Canadian surveillance programs (2005-2010). Foodborne Pathogens and Disease. 2013;**10**:747-756. DOI: 10.1089/fpd.2012.1438

[122] Calistri P, Iannetti S, Danzetta L, Narcisi M, Cito V, Di Sabatino F, et al. The components of 'OneWorld-one health' approach. Transboundary and Emerging Diseases. 2013;**60**(Suppl 2): 4-13. DOI: 10.1111/tbed.12145

[123] Wendt A, Kreienbrock L, Campe A. Zoonotic disease surveillance – Inventory of systems integrating human and animal disease information. Zoonoses and Public Health. 2015;**62**(1):61-74. DOI: 10.1111/zph.12120

[124] Carmena D, Cardona GA. Echinococcosis in wild carnivorous species: Epidemiology, genotypic diversity, and implications for veterinary public health. Veterinary Parasitology. 2014;**202**:69-94. DOI: 10.1016/j.vetpar.2014.03.009

[125] Weldhagen GF. Integrons and β-lactamases- a novel perspective on resistance. International Journal of Antimicrobial Agents. 2004;**23**(6):556-562. DOI: 10.1016/j.ijantimicag.2004.03.007

[126] Zhou H, Zhang T, Yu D, Pi B, Yang Q, Zhou J, et al. Genomic analysis of the multidrug-resistant *Acinetobacter baumannii* strain MDR-ZJ06 widely spread in China. Antimicrobial Agents and Chemotherapy. 2011;**55**(10):4506-4512. DOI: 10.1128/AAC.01134-10

[127] Fallah F, Taherpour A, Vala MH, Hashemi A. Global spread of New Delhi mettallo-β-lactamase-1 (NDM-1). Archives of Clinical Infectious Diseases. 2012;**6**(4):171-177. DOI: 10.1155/2014/245162

[128] Bonnin RA, Potron A, Poirel L, Lecuyer H, Neri R, Nordmann P. PER-7, an extended-spectrum β-lactamase with increased activity toward broad-spectrum cephalosporins in *Acinetobacter baumannii*. Antimicrobial Agents and Chemotherapy. 2011;**55**(5):2424-2427. DOI: 10.1128/AAC.01795-10

[129] Santiago GS, Coelho IS, Moreira A, Farias BO, Alencar TA, Souza MMS, et al. Detection of mutations in ampC promoter/attenuator gene in *Escherichia coli* from dairy cows in Rio de Janeiro and Mato Grosso, Brazil. African Journal of Microbiology Research. 2019;**13**(25):388-391. DOI: 10.5897/AJMR2019.9134

[130] Khoshnood S, Heidary M, Mirnejad R, Bahramian A, Sedighi M, Mirzaei H. Drug-resistant gram-negative uropathogens: A review. Biomedicine & Pharmacotherapy. 2017;**94**:982-994. DOI: 10.1016/j.biopha.2017.08.006

[131] Bush K, Jacoby GA. Updated functional classification of β-lactamases. Antimicrobial Agents and Chemotherapy. 2010;**54**(3):969-976. DOI: 10.1128/AAC.01009-09

[132] Ambler RP. The structure of beta-lactamases. Philosophical Transactions of the Royal Society B. 1980;**289**(1036):321-331. DOI: 10.1098/rstb.1980.0049

[133] Tooke CL, Hinchliffe P, Bragginton EC, Colenso CK, Hirvonen HA, Takebayashi Y, et al. β-Lactamases and β-lactamase inhibitors in the 21st century. Journal of Molecular Biology. 2019;**431**(18):3472-3500. DOI: 10.1016/j.jmb.2019.04.002

[134] Ghafourian S, Sadeghifard N, Soheili S, Sekawi Z. Extended Spectrum Beta-lactamases: Definition, classification and epidemiology. Current Issues in Molecular Biology. 2015;**17**:11-21

[135] Paterson DL, Bonomo RA. Extended-spectrum β-lactamases: A clinical update. Clinical Microbiology Reviews. 2005;**18**:657-686. DOI: 10.1128/CMR.18.4.657-686.2005

[136] Knothe H, Shah P, Krcmery V, Antal M, Mitsuhashi S. Transferable resistance to cefotaxime, cefoxitin, cefamandole and cefuroxime in clinical isolates of *Klebsiella pneumoniae* and *Serratia marcescens*. Infection. 1983;**11**:315-317. DOI: 10.1007/bf01641355

[137] Brun-Buisson C, Legrand P, Philippon A, Montravers F, Ansquer M, Duval J. Transferable enzymatic resistance to third-generation cephalosporins during nosocomial outbreak of multiresistant *Klebsiella pneumoniae*. Lancet. 1987;**2**:302-306. DOI: 10.1016/s0140-6736(87)90891-9

[138] Philippon A, Labia R, Jacoby G. Extended-spectrum β-lactamases. Antimicrobial Agents and Chemotherapy. 1989;**33**:1131-1136. DOI: 10.1128/aac.33.8.1131

[139] Bauernfeind A, Grimm H, Schweighart S. A new plasmidic cefotaximase in a clinical isolate of *Escherichia coli*. Infection. 1990;**18**:294-298. DOI: 10.1007/bf01647010

[140] Rivoarilala OL, Garin B, Andriamahery F, Collard JM. Rapid in vitro detection of CTXM groups 1, 2, 8, 9 resistance genes by LAMP assays. PLoS One. 2018;**13**(7):e0200421. DOI: 10.1371/journal.pone.0200421

[141] Silva KC, Lincopan N. Epidemiology of extended-spectrum β-lactamases in Brazil: Clinical impact and implications for agribusiness. Jornal Brasileiro de Patologia e Medicina Laboratorial. 2012;**48**(2):91-99

[142] Radice M et al. Early dissemination of CTX-M-derived enzymes in South America. Antimicrobial Agents and Chemotherapy. 2002;**46**(2):602-604. DOI: 10.1590/S1676-24442012000200004

[143] Minarini LAR et al. Clonal transmission of ESBL- producing *Klebsiella* spp. at a university hospital in Brazil. Current Microbiology. 2008;**56**(6):587-591. DOI: 10.1007/s00284-008-9129-5

[144] Sartori L, Sellera P, Moura Q, Cardoso B, Cerdeira L, Lincopan N. Multidrug-resistant CTX-M-15-positive *Klebsiella pneumoniae* ST307 causing urinary tract infection in a dog in Brazil. Journal of Global Antimicrobial

Resistance. 2019;**19**:96-97. DOI: 10.1016/j.jgar.2019.09.003

[145] Kohlmann R, Nefedev A, Kaase M, Gatermann SG. Community-acquired adult *Escherichia coli* meningitis leading to diagnosis of unrecognized retropharyngeal abscess and cervical spondylodiscitis: A case report. BMC Infectious Diseases. 2018;**15**:567. DOI: 10.1186/s12879-015-1310-4

[146] Mariappan S, Sekar U, Kamalanathan A. Carbapenemase-producing Enterobacteriaceae: Risk factors for infection and impact of resistance on outcomes. International Journal of Applied & Basic Medical Research. 2017;**7**(1):32-39. DOI: 10.4103/2229-516X.198520

[147] Decousser JW, Poirel L, Nordmann P. Recent advances in biochemical and molecular diagnostics for the rapid detection of antibiotic-resistant Enterobacteriaceae: A focus on ss-lactam resistance. Expert Review of Molecular Diagnostics. 2017;**17**:327-350. DOI: 10.1080/14737159.2017.1289087

[148] European Antimicrobial Susceptibility Testing Committee. Guidelines for Detection of Resistance Mechanisms and Specific Resistances of Clinical and/or Epidemiological Importance. The European Committee on Antimicrobial Susceptibility Testing. 2017. Available from: http://www.eucast.org.

[149] Santiago GS, Motta CC, Almeida GFB, Goncalves D, Souza MMS, Coelho IS, et al. A review: AmpC β-lactamase production in Enterobacteriaceae. The Brazilian Journal of Veterinary Medicine. 2016;**38**:17-30

Chapter 9

Scenario of Antibiotic Resistance in Developing Countries

Mohammad Mahmudul Hassan

Abstract

Antibiotic resistance is an emerging global concern. It is an increasing threat to public health sectors throughout the world. This devastating problem has drawn attention to researchers and stakeholders after a substantial economic loss for decades resulting from the ineffectiveness of antibiotics to cure infectious diseases in humans and animals. The spectrum of antibiotic resistance varies between developed and developing countries due to having variations in treatment approaches. Antibiotic therapy in the developed countries is usually rational and targeted to specific bacteria, whereas in the developing countries, most of the cases, the use of antibiotics is indiscriminate to the disease etiology. In developing countries, many people are not aware of using antimicrobials. They usually get suggestions from drug sellers and quacks who do not have the authorization to prescribe a drug. If registered doctors and veterinarians are asked to prescribe, then dose, course, and withdrawal period might be maintained adequately. Antibiotic resistance transmission mechanisms between agricultural production systems, environment, and humans in developing countries are very complex. Recent research makes a window to find out the global situation of antibiotic use and resistance pattern. The antibiotic resistance scenario in selected developing countries has been summarized in this chapter based on published literature (**Table 1**). This chapter describes the judicial use of antibiotics and discussed maintaining proper antibiotic dose, course, drug withdrawal period, especially on food-producing animals. The book contains a few recommendations, suggested by the national multi-sectoral surveillance committee to avoid antibiotic resistance organisms in livestock and humans in the developing countries.

Keywords: Antibiotics, Antibiotic resistance pattern, prescribed, registered doctors, developing countries

1. Introduction

After discovering the first antibiotic 'Penicilillin' by Alexander Fleming in 1928, antibiotics played a notable role in saving millions of lives globally. Nowadays, the resistance of antibiotics has intensified significantly throughout the world [1]. Antibiotic resistance is a global problem in both developed and developing countries. The incidence of resistance has increased at an alarming rate in recent years and is expected to increase at a greater rate in the future as antibiotic agents continue to lose their efficiency [2], mostly in many developing or low-and middle-income countries (LMIC). Resistance bacteria do not respect national borders; the development of resistance in the most remote locations can impact the world in a concise time [1]. The widespread use of antibiotics for human and veterinary treatment has led to large-scale dissemination of bacteria with resistance ability to antibiotics in the domestic animal-wildlife-environmental niche via food chain to humans in most developing countries, including Bangladesh [3]. Resistance bacteria are found in the stool and as intestinal flora of healthy individuals that are serving as reservoirs for resistance to multiple antimicrobials [4]. Antibiotics are a mainstay in the treatment of bacterial infections, and thus the worldwide increase in antibiotic-resistance bacteria is of major concern. The problem of antibiotic resistance is not restricted to pathogenic bacteria—it also involves the commensal microbiota, which may become a major reservoir of resistance strains of bacteria [5]. *Escherichia coli* is commonly found in the intestinal tract of humans and animals and can also be concerned with human and animal infectious diseases. Animal food products are important sources of *E. coli* as fecal contamination of processed animal carcasses at the slaughterhouse is frequently occurred. These resistance microorganisms and their possible resistance determinants may be transmitted to humans if these animal origin foods are improperly washed, cooked, or otherwise mishandled [6]. Although most isolates of *E. coli* are nonpathogenic, they are considered an indicator of fecal contamination in food. About 10 to 15% of intestinal coliforms are opportunistic and pathogenic serotypes and cause a variety of lesions in immunocompromised hosts such as animals and humans [7]. Among the diseases that they cause, some are often severe and sometimes lethal such as- meningitis, endocarditis, urinary tract infection, septicemia, and epidemic diarrhea in human, and yolk sac infection, omphalitis, cellulitis, swollen head syndrome, coligranuloma, and colibacillosis in birds [8]. Furthermore, salmonellosis is one of the most frequent foodborne diseases in humans in almost all countries, and *Salmonella enterica ssp. enteritidis*, followed by *typhimurium*, represent the most frequently isolated serotypes [9]. The most common disease syndromes caused by *Salmonella* serotypes in humans are typhoid fever and enteritis [10], and in avian species, *Salmonella* organism causes fowl typhoid and pullorum disease [11]. *Salmonella typhimurium* and *S. dublin* appear to be the commonest serotypes isolated from cattle, although the distribution of these 2 serotypes differs between countries, and the *Salmonella* organism predominantly causes bovine salmonellosis [12]. *S. aureus* causes superficial skin lesions and localized abscesses in a wide range of host animals. *S. aureus* causes deep-seated infections, such as osteomyelitis and endocarditis and more serious skin infections [13]. *S. aureus* is a major cause of hospital-acquired (nosocomial) infection of surgical wounds and, with *S. epidermidis*, causes infections associated with indwelling medical devices [14]. It also causes food poisoning by releasing enterotoxins into animal originated food. *S. aureus* causes toxic shock syndrome by release of superantigens into the blood stream. *S. saprophiticus* causes urinary tract infections in human, frequently in female population [15]. Over the past decade, the changing pattern of resistance against bacteria has depicted the need for new antimicrobial agents [2]. Developing countries are more vulnerable to antimicrobial resistnace issues for their underprivileged health care infrastructure, unregulated agricultural production process, poor sanitation facilities and widespread misuse of antibiotics. In addition, weak monitoring system and improper implimentation of legislative practices on antibiotic sell and uses in the agriculrural production systems, increases the possibilities of registant bacteria in the developing countries. The senario of antibiotic resistance pattern worsen in developing countries as they use antibiotic indiscriminately in clinical treatments and food animal production system as well. With many bacterial causing diseases in human and animal in developing countries, this chapter will be focusing on three most common genera of bacteria viz. Escherichia, *Salmonella* and *Staphylococcus* that are posing threat to public health by gradually getting resistance against many antibiotics. The aim of this chapter is to identify the scenario of antibiotic resistance pattern in developing countries based on published literature (**Table 1**) and compile them to find out the overall spectrum of antibiotic resistance.

Country	Host	Bacteria			Author
		Escherichia coli	*Salmonella* spp.	*Staphylococcus* spp.	
Pakistan	Human	Amoxicillin, Ampicillin, Aztreonam, Cephalosporin, Cefotaxime, Ceftriaxone Ciprofloxacin, Floroquinol, Trimethoprim-sulfamethoxazole		Amoxicillin, Ampicillin, Amikacin, Cefoxitin, Chloramphenicol, Ciprofloxacin, Clindamycin, Co-trimoxazole, Doxycycline, Erythromycin, Fusidic acid, Gentamicin, Penicillin, Teicoplanin, Tigecycline, Levofloxacin, Linezolid, Vancomycin	[16–18]
	Poultry and poultry products	Ampicillin, Ciprofloxacin, Colistin, Tetracycline	Pefloxacin	Cefoxitin, Gentamicin, Oxacillin, Penicillin, Levofloxacin, Moxifloxacin	[19–22]
	Livestock			Amikacin, Amoxicillin, Cefoxitin, Ampicillin, Oxacillin, Augmentin, Cefotaxime, Chloramphenicol, Ciprofloxacin, Clindamycin, Enrofloxacin, Erythromycin, Fosfomycin, Gentamycin, Kanamycin, Linezolid, Ofloxacin, Penicillin, Rifampicin, Teicoplanin, Trimethoprim, Vancomycin	[23, 24]
India	Human	Amikacin, Ampicillin, Ampicillin, Augmentin, Cefepime, Cefoxitin, Cefoperazone, Cefotaxime, Quinolones, Ceftazidime, Ceftriaxone, Colistin, Cefuroxime, Cephalosporins, Ciprofloxacin, Co-trimoxazole, Ertapenem, Meropenem, Gentamycin, Imipenem, Nalidixic acid, Nitrofurantoin, Norfloxacin, Piperacillin, Streptomycin, Sulfamethoxazole, Tetracycline	Ampicillin, Azithromycin, Ceftriaxone, Cephalosporins, Chloramphenicol, Fluoroquinolones, Trimethoprim	Ciprofloxacin, Clindamycin, Co-trimoxazole, Erythromycin, Gentamicin	[25–35]
	Poultry and poultry products	Amoxicillin, Ampicillin, Cephalexin, Colistin, Cefoxitin, Chloramphenicol, Neomycin, Ciprofloxacin, Co-trimoxazole, Trimethoprim, Amoxicillin, Erythromycin, Rifamycin, Streptomycin, Doxycycline, Sulfamethoxazole, Nalidixic acid, Tetracycline, Gentamicin	Sulphamethizole, Chloramphenicol Amikacin Ceftazidime, Oxytetracycline, Nalidixic acid	Penicillin, Ciprofloxacin, Tetracycline, Erythromycin Ampicillin, Tetracycline Amoxicillin, Erythromycin Polymyxin-B, Cefoxitin Novobiocin, Oxacillin	[36–47]

Country	Host	Bacteria			Author
		Escherichia coli	***Salmonella* spp.**	***Staphylococcus* spp.**	
	Livestock	Meropenem, Imipenem, Ertapenem		Methicillin, Penicillin, Ampicillin, Kanamycin, Cefotaxime, Sulphadizine Amoxicillin	[48–51]
	Pet animals			Amoxicillin, Penicillin G, Methicillin, Cloxacillin, Ampicillin, Cephalothin, Cefuroxime, Ceftriaxone, Clavulanate, Neomycin, Streptomycin, Furazolidone, Nitrofurantoin, Ciprofloxacin, Erythromycin, Oleandomycin, Azithromycin, Doripenem, Lincomycin, Clindamycin, Sulfafurazole, Sulfadiazine, Chloramphenicol, Novobiocin, Vancomycin	[50, 52]
	Food and food products	Colistin, Cefotaxime, Ceftazidime, Gentamicin, Tetracycline, Amoxicillin		Oxacillin, Cefoxitin, Penicillin G, Cephalexin, Ampicillin, Methicillin, Kanamycin, Gatifloxacin, Ciprofloxacin	[53–56]
	Environment	Amoxicillin, Ciprofloxacin, Nalidixic acid, Ceftazidime, Cephalosporin, Penicillin, Cefuroxime, Erythromycin, Tetracycline, Ceftazidime, Cefotaxime, Gentamicin, Trimethoprim			[57–60]
Bangladesh	Human	Colistin, Nalidixic Acid, Cefixime, Co-trimoxazole, Ceftazidime, Gentamicin, Amikacin, Imipenam, Ciprofloxacin, Azithromycin, Cefuroxime, Cefotaxime, Ceftriaxone, Meropenem, Nitrofurantoin, Levofloxacin, Meropenem	Ciprofloxacin, Ceftriaxone, Azithromycin, Clindamycin	Nalidixic Acid, Cefixime, Meropenem, Oxacillin, Gentamicin, Ceftazimid, Tocefoxitin, Etracylcin, Cefoxitin, Ciprofloxacin, Chloramphenicol, Clindamycin, Cefotaxime, Levofloxacin	[13, 14, 61–68]
	Poultry	Ampicillin, Tetracycline, Trimethoprim, Nalidixic acid			[7]

Country	Host	Bacteria			Author
		Escherichia coli	*Salmonella* spp.	*Staphylococcus* spp.	
	Food and food products	Erythromycin, penicillin, Vancomycin, Novobiocin, Tetracycline, Ceftriaxone, Ampicillin, Azithromycin, Bacitracin, Kanamycin, Nalidixic acid, Sulfamethoxazole	Ampicillin, Azithromycin, Erythromycin, Doxycycline, Sulphonamide, Azithromycin, Novobiocin, Oxytetracycline, Cephradine, Amoxicillin, Erythromycin, Tetracycline, Erythromycin, Vancomycin, Rifampicin, Sulfamethoxazole, Bacitracin	Ampicillin, Chloramphenicol, Nitrofurantoin, Oxytetracycline, Oxytetracycline, Amikacin, Erythromycin, Oxacillin, Ciprofloxacin, Amoxicillin, Trimethoprim, Gentamicin, Penicillin, Erythromycin	[69–74]
	Environment	Ceftazidime, Gentamycin, Tetracycline, Imipenem, Ciprofloxacin, Chloramphenicol, Amoxycillin, Erythromycin, Azithromycin, Streptomycin, Norfloxacin, Cefepime, Cefixime	Ceftazidime, Gentamycin, Imipenem, Ciprofloxacin, Chloramphenicol, Cefoxitin, Tetracycline, Rifampicin, Ampicillin	Ceftazidime, Gentamycin, Azithromycin, Tetracycline, Imipenem, Ciprofloxacin, Chloramphenicol, Methicillin, Vancomycin	[75–77]
Thailand	Human	Trimethoprim/sulfamethoxazole, Colistin, Amoxicillin, Gentamicin, Cefazolin, Cefotaxime, Ceftazidime, Ceftriaxone, Cefixime, Cefalexin, Nalidixic acid, Ciprofloxacin, Norfloxacin, Ofloxacin, Doxycycline, Nitrofurantoin, Ampicillin, Oxacillin, Amikacin, Aztreonam, Cefotaxime, Meropenem, Piperacillin, Chloramphenicol, Amoxycillin, Cotrimoxazole	Trimethoprim-Sulfamethoxazole, Cefotaxime, Ceftazidime, Ceftriaxone, Ceftazidime, Norfloxacin, Nalidixic acid, Tetracycline, Gentamicin, Ampicillin, Ciprofloxacin, Chloramphenicol, Cotrimoxazole	Fosfomycin, Methicillin, Cefoxitin, Penicillin, Oxacillin, Mupirocin, Rifampicin, Cotrimoxazole, Ciprofloxacin, Chloramphenicol, Cefazolin, Clindamycin, Cephalexin, Trimethoprim, Amikacin, Ampicillin, Amoxicillin, Tetracycline, Cloxacillin, Cefotaxime, Meropenem, Piperacillin, Gentamicin, Ofloxacin, Erythromycin	[78–87]
	Livestock			Methicillin, Penicillin, Rifampin, Novobiocin, Tetracycline, Clindamycin, Oxacillin, Linezolid, Erythromycin, Cefoxitin, Kanamycin, Gentamicin, Trimethoprim, Ciprofloxacin, Levofloxacin	[88]
	Food and food products	Ampicillin, Cefotaxime, Cefpodoxime, Aztreonam, Ceftazidime, Imipenem, Gentamicin, Amoxicillin, Ceftriaxone, Nalidixic acid, Amoxicillin, Ampicillin, Cefepime, Amikacin, Doxycycline, Tetracycline, Ciprofloxacin, Co-trimoxazole, Colistin sulfate, Cefoxitin, Enrofloxacin,			[89, 90]

Country	Host	Bacteria			Author
		Escherichia coli	*Salmonella* spp.	*Staphylococcus* spp.	
		Erythromycin, Chloramphenicol, Ceftazidime, Trimethoprim			
	Environment	Penicillin G, Vancomycin, Erythromycin, Ampicillin, Tetracycline, Chloramphenicol, Streptomycin, Neomycin, Kanamycin, Cefazoline, Cefotaxime, Ceftazidime, Gentamicin, Nalidixic acid	Tetracycline, Ampicillin, Streptomycin, Tetracycline, Trimethoprim, Gentamicin, Ciprofloxacin, Nalidixic acid, Penicillin G, Neomycin, Vancomycin, Erythromycin, Kanamycin, Chloramphenicol	Methicillin	[91–94]
Nepal	Human	Amikacin, Ampicillin, Cefotaxime, Levofloxacin, Ciprofloxacin, Gentamicin, Ampicillin, Cefoxitin, Trimethoprim, Nitrofurantoin, Amoxyclav, Piperacillin, Ofloxacin, Cefotaxime, Colistin, Meropenem, Nitrofurantoin, Norfloxacin, Imipenem, Fosfomycin, Cefixime, Piperacillin, Cefoperazone, Nitrofurantoin, Meropenem, Co-trimoxazole, Ceftriaxone, Levofloxacin, Ceftazidime, Chloramphenicol, Nalidixic acid, Piperacillin, Tetracycline	Ciprofloxacin, Ampicillin, Chloramphenicol, Co-trimoxazole, Streptomycin, Nalidixic acid, Trimethoprim-Sulfamethoxazole, Ceftriaxone	Ampicillin, Ceftriaxone, Cefotaxime, Cefixime, Nalidixic acid, Piperacillin, Penicillin, Erythromycin, Clindamycin, Cefoxitin, Chloramphenicol, Ampicillin, Ciprofloxacin, Cotrimoxazole, Cefoxitin, Gentamicin, Tetracycline, Teicoplanin, Cephalexin, Cloxacillin, Erythromycin, Linezolid, Vancomycin, Ampicillin, Azithromycin	[15, 95–107]
	Poultry	Ampicillin, Amikacin			[108]
	Food and food products	Amoxicillin, Tetracycline, Cefotaxime, Nalidixic acid, Cotrimoxazole, Gentamycin	Tetracycline, Chloramphenicol, Nalidixic acid, Amoxicillin	Amoxicillin, Nalidixic acid, Cefotaxime, Tetracycline, Azithromycin, Cotrimoxazole	[109, 110]
Nigeria	Human	Cefuroxime, Cefotaxime, Amoxicillin, Imipenem, Ceftriaxone, Cefalexin, Ampicillin, Ciprofloxacin, Nalidixic Acid, Gentamycin, Nitrofurantoin, Kanamycin, Chloramphenicol, Pefloxacin, Ofloxacin, Streptomycin, Ceftazidime, Tetracycline, Amoxicillin, Trimethoprim, Co-trimoxazole	Ampicillin, Cefotaxime, Chloramphenicol, Trimethoprim-sulfamethoxazole, Ofloxacin, Ciprofloxacin, Co-trimoxazole, Tetracycline, Eftazidime, Ceftriaxone	Streptomycin, Gentamycin, Tetracycline, Cotrimoxazole, Erythromycin, Cloxacillin, Chloramphenicol, Augmentin, Imipenem, Ceftriaxone, Cefoxitin, Ciprofloxacin, Erythromycin, Cefalexin Co-trimoxazole, Nalidixic Acid, Ampicillin, Vancomycin, Azithromycin, Cefuroxime, Amoxicillin, Ceftazidime	[111–120]

DOI: http://dx.doi.org/10.5772/intechopen.94957

Country	Host	Bacteria			Author
		Escherichia coli	***Salmonella* spp.**	***Staphylococcus* spp.**	
	Poultry	Tetracycline, Ampicillin Nitrofurantoin, Chloramphenicol, Penicillin, Ampicillin, Amoxicillin, Cloxacillin, Augmentin, Tetracycline, Streptomycin, Gentamicin, Erythromycin, Cotrimoxazole, Nalidixic Acid		Penicillin, Ampicillin, Amoxicillin, Cloxacillin, Augmentin, Tetracycline, Streptomycin, Gentamicin Chloramphenicol, Ofloxacin, Erythromycin, Cefuroxime, Cefoxitin, Amoxicillin, Ceftriaxone	[121–123]
	Livestock	Cloxacillin, Penicillin, Teicoplanin, Sulphadimidine, Ampicillin, Tetracycline, Nalidixic acid, Trimethoprim	Amoxicillin, Enrofloxacin	Ampicillin	[124, 125]
	Environment	Ceftazidime, Cephalexin, Ceftriaxone, Cefotaxime, Cephalexin, Tetracycline, Lipocaine, Augmentin, Ceftazidime, Cefuroxime, Ampicillin, Chloramphenicol, Amoxicillin-clavulanic acid, Ciprofloxacin, Ampicillin, Augmentin, Gentamicin	Gentamycin, Ofloxacillin, Amoxycillin, Ciprofloxacin, Tetracycline, Pefloxacin, Lipocaine, Ceftazidime, Ceftriaxone, Cefotaxime, Cefotaxine, Cephalexin, Augmentin, Cefuroxime, Ampicillin, Colistin, Ofloxacin, Cotrimoxazole, Ciprofloxacin, Nitrofurantoin Trimethoprim, Ceftazidime	Ceftazidime, Cephalexin, Ceftriaxone, Cephalexin, Tetracycline, Lipocaine, Amoxicillin	[126–132]
Brazil	Poultry		Amoxicillin, Ceftiofur, Ciprofloxacin, Gentamicin, Chloramphenicol, Tetracycline, Sulfafurazole, Enrofloxacin, Sulfonamide, Spectinomycin, Trimethoprim		[133, 134]
	Human		Ampicillin, Ampicillin, Ceftriaxone, Ceftiofur, Chloramphenicol, Ciprofloxacin, Enrofloxacin Tetracycline, Trimethoprim		[135]
	Food and food products		Sulfonamides, Streptomycin, Tetracycline, Gentamicin, Ceftriaxone, Trimethoprim		[136, 137]

Table 1.
Summary of antibiotic resistance scenario of three bacteria in different samples from selected developing countries.

2. Main text

2.1 Practical scenario of antibiotic resistance pattern in developing countries

An organized literature search approach was used to detect all published studies reporting resistance bacteria in human samples and foods of animal origin in some selected developing countries. PubMed, Science Direct, and Google Scholar were searched for relevant studies published until 2019. The search terms have been adopted into outcome, population, descriptive, and area categories. Based on the study objectives, specific Boolean words were developed using "AND" and "OR". Some modification has been conducted based on the search engine requirements, and advanced search criteria have been used to search Google scholar. The papers were downloaded using the Chattogram Veterinary and Animal Sciences University (CVASU) library network. The Boolean words of each category were combined using "AND", whereas "OR" was used to join the term within a category. Data was extracted and recorded for study location, citation, first author, title, time of study, year of publication, type of specimen, sample size, number of positive specimens, amount of antibiotics, specific antibiotic sensitivity or resistance level percentages, methods of detection used, culturing techniques and resistance genes. Resistance of *E. coli* was mostly seen in humans and poultry compared to *Salmonella* and *Staphylococcus*, and the most resistance drug was Ampicillin and Ciprofloxacin in Pakistan. Furthermore, resistance of salmonella was seen in human samples with Ampicillin, Trimethoprim, and Ceftriaxone. Pefloxacin was resistance to *Salmonella* in derived from poultry. Resistance staphylococcus were observed in cattle, buffalo, poultry, and table egg to antibiotics Penicillin, Ampicillin, Oxacillin, Ciprofloxacin, Trimethoprim, Gentamicin, Linezolid, Erythromycin, Clindamycin, Amikacin, Vancomycin, Chloramphenicol and Cefoxitin. In India, resistance of *E. coli* was mostly seen in poultry, and the human was in second position and the drugs: Ciprofloxacin, Ampicillin, Amoxicillin, Trimethoprim, Gentamicin, Co-trimoxazole and Sulfamethoxazole were found resistance. The highest resistance of Salmonella was detected in poultry with a higher level of Oxytetracycline. In the case of *Staphylococcus* spp., excessive resistance was seen in poultry and cattle with commonly used antimicrobials: Oxacillin, Penicillin G, Ampicillin, Methicillin, Amoxicillin, Erythromycin, Methicillin, Cloxacillin, and Kanamycin. In Bangladesh, the highest antibiotic resistance of *E. coli* was seen in human, and the most resistance drugs are Tetracycline, Ampicillin, Nalidixic acid, Trimethoprim-Sulfamethoxazole, Ciprofloxacin, and Ceftriaxone. Moreover, *Salmonella* resistance to Azithromycin, Ampicillin, and Erythromycin was detected in humans. Resistance of *Staphylococcus* was observed in humans, and the most resistance antibiotics are Ciprofloxacin, Gentamicin, Chloramphenicol, Tetracycline, and doxycycline. In Thailand, the highest resistance of *E. coli* was noticed in human and pig, and the most resistance antibiotics are Ampicillin, Ceftazidime, Tetracycline, Gentamicin, Ciprofloxacin, Norfloxacin, Clavulanic acid, Doxycycline and Colistin sulfate. Research revealed that resistance *Salmonella* was detected in the Thai human population alongside highly resistance antibiotics: Ampicillin, Tetracycline, Ciprofloxacin, Chloramphenicol, and Trimethoprim. On the other hand, resistance *Staphylococcus* was found in humans with higher drug resistance, and the antibiotics were Doxycycline, Gentamicin, Cefoxitin, Ceftriaxone, Methicillin, Tetracycline, Erythromycin, Penicillin, and Cefoxitin. In Nepal, higher resistance of *E. coli* was identified in humans, and many bacteria became resistance, including Doxycycline, Gentamicin, Cefoxitin, Ceftriaxone, Methicillin, Tetracycline, Erythromycin, Penicillin, and Cefoxitin. Besides, resistance salmonella was recognized in humans and foods with resistance antibiotics such as Ampicillin, Ciprofloxacin, Chloramphenicol,

Co-trimoxazole, Nalidixic acid, and Amoxicillin. However, antibiotics such as Amikacin, Gentamicin, Ciprofloxacin, Amoxicillin, Tetracycline, Erythromycin, Cefotaxime, Oxacillin, Cefoxitin and Co-trimoxazole recorded resistance against Staphylococcus in Nepal. In Nigeria, the highest resistance of *E. coli* was reported in human and resistance antibiotics were Tetracycline, Ceftazidime, Cefotaxime, Ceftriaxone, Ciprofloxacin, Gentamycin, Sulfamethoxazole, Penicillin, Ampicillin, Amoxicillin, Cloxacillin, Augmentin and Amoxicillin. Moreover, resistance Salmonella was found in the water source in the environment to antibiotics Ampicillin, Cefotaxime, Ceftazidime, Ciprofloxacin, Sulfamethoxazole-trimethoprim, and Tetracycline. Moreover, the resistance *Staphylococcus* was seen in humans and the environment, and the resistance antibiotics were Ceftriaxone, Gentamicin, Erythromycin, Co-trimoxazole, Chloramphenicol, Tetracycline, Streptomycin, Cephalexin, and Ampicillin. Finally, in Brazil, antimicrobial-resistance (AMR) *E. coli* were recorded in water source, and the resistance antibiotics were Ampicillin, Cephalexin, Amoxicillin, and Polymyxin. On the other hand, resistance salmonella was detected in poultry with resistance antibiotics such as Gentamicin, Sulfonamide, Trimethoprim, Ampicillin, and Chloramphenicol, Ciprofloxacin, Enrofloxacin, Tetracycline, and Ceftriaxone. A great majority of antimicrobial classes that are already resistance to the bacteria are used in humans and animals, including domestic animals, poultry and other birds, and commercial farm fishes. These findings of AMR in the agricultural production system, environment, and humans from developing countries pose a threat to the global context.

2.2 Tale of AMR in developing countries

Antibiotics are considered to safeguard against infectious diseases caused by pathogenic bacteria, but unfortunately, antimicrobial resistance becomes a burden in humans, animals, and the environmental niche worldwide. It happened due to the indiscriminate, inappropriate, and unregulated use of antibiotics in animal and agricultural production systems and humans. In developing countries, AMR is overburdened by antibiotics as growth promoters by the farmers, feed dealers, drug sellers, and the lack of approved legislation by the respective government authorities [138]. However, some countries have written and approved legislation, but appropriate implementation and systematic monitoring are not noticed. Multi-drug resistance (MDR) bacteria are increasing day by day at every corner of developing countries and escalate treatment costs. In a recent WHO report, it is speculated that about 10 million people will die, and 100 trillion USD from the world economy will be lost for AMR by 2050 if no effective measures are taken [139]. Humans are mostly suffering in developing countries due to the ineffectiveness of antibiotics to microbes. *E. coli*, *Salmonella* spp. and *Staphylococcus* spp. are now resistance to the commonly used antibiotics and some higher generation antibiotics such as 3rd generation cephalosporins. This might be due to cross-contamination with hospital equipment, animal originated food, and mixing of medical and veterinary hospital effluents in the environments [16, 26, 31, 67, 78, 97].

In highly populated developing countries where there is a shortage of physicians, the people seek to take drugs, including antibiotics, by their own decision or prescription from drug sellers or quacks. Even in the rural area, it is hard to find a licensed doctor or veterinarian to treat people and animals and keep faith in a quack or village doctor. Those quacks, health assistant village doctors, and drug sellers prescribe different antibiotics even for common symptoms such as colds, coughs, and diarrhea, where a simple, supportive treatment course would be enough. Self-medication, both in the human and veterinary sectors, is another major problem for generating antimicrobial resistance. In some cases, licensed doctors and veterinarians are biased to treat antimicrobials due to various pharmaceutical companies [138]. Those unnecessary prescriptions and a broad spectrum of antibiotics in animals and humans have already brought a great disaster in most developing countries [29]. Poor sanitation and hygiene are essential factors for transmitting resistance organisms from animals (mainly food and pet animals) and environment to humans. Countries like Bangladesh, Brazil, India, and Nigeria are mostly suffering from sanitation and hygiene management issues for growing AMR [140]. There is a chance of nosocomial infection in hospital settings, as many hospitals have no facilities for waste disposal and wastewater treatment [14]. There is also a high risk of spreading resistance microbes from patients to their surroundings, especially to caregivers or family members.

Poultry meat is one of the topmost widely accepted food worldwide as a cheap protein source, and more than 90 billion tons of chicken meat produce each year. A large variety of antimicrobials are used in poultry production systems for disease prophylaxis and used as growth promoters to increase growth and productivity [8], which accelerate the expansion of resistance in pathogens and different commensals. Therefore, human health is a great concern with the emergence of resistance pathogens from poultry and AMR residue from poultry meat and eggs [18, 74]. Food producing animals or livestock has, also affected by AMR due to not maintaining proper dose, treatment interval and duration in therapeutics, metaphylactic and prophylactic treatment, and withdrawal periods of different antimicrobials. Growth promoter is another influential factor-like poultry production system in most developing countries [88, 124, 135]. Human-livestock interaction is another vital factor for transmitting resistance microorganisms from food and pet animals to humans or vice versa.

An agreement should be maintained among the scientific community to stop the excessive use of antimicrobials in food animal production system. Thus, it will help to limit the AMR on human health. Otherwise, AMR in food animal pathogens will unavoidably effect on treatment failure of livestock and poultry diseases. As a result, pathogen transmission on the environment will increase, and production loss will be soared, and the economy of developing countries will be hindered. In developing countries, the environment is also contaminated with high levels of resistance organisms and AM residues derived from human, livestock, and poultry waste [124]. Hospital, both human and veterinary wastewater, is the potential source of AMR.

Water is the mainstream potential reservoir of antimicrobial resistance as wastewater contaminated rivers, ponds, and other water bodies. Medical and veterinary hospital effluents (with different types of resistance organisms) were directly drained to the nearby water bodies and contaminated the fishes ultimately consumed by humans. Poor sanitation and hygiene management bring pathogens close to each other's species and accelerate the horizontal resistance gene transmission [140]. Ceftazidime, Cefpodoxime-resistance bacteria were isolated in Nigeria. Moreover, Azithromycin, Tetracycline, Gentamicin, Ciprofloxacin, Cefotaxime, Chloramphenicol, Cefoxitin, and Oxacillin resistance *Staphylococcus aureus* found in both human and veterinary hospital drainage water in Bangladesh [14, 121]. Research in Thailand detected Cefazoline, Cefotaxime, Ceftazidime, Gentamicin, Tetracycline, Chloramphenicol, Kanamycin, and Nalidixic acid resistance *E. coli*, which indicate the vulnerability of AMR in the environment [94]. In food animals in developing countries, antibiotics are frequently used in food and water to the entire group for a prolonged time and often at sub-therapeutic doses. These conditions favor the selection and spread of resistance bacteria within and between animals as well as to humans through food consumption and other environmental pathways.

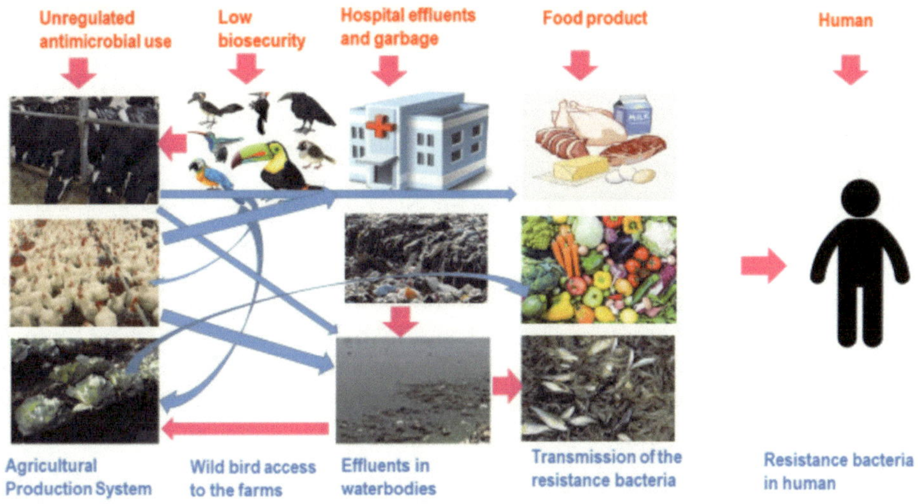

Figure 1.
Complex transmission dynamics of AMR between agricultural production system, environment, and human (credit: MM Hassan; created by using online materials).

To reduce the AMR in developing countries, proper rules and regulations for antibiotic use in humans and animals should be followed. Only registered physicians will prescribe antibiotics for humans; livestock and poultry farming will be conducted with veterinary supervision. Buying and selling antimicrobials should be restricted without prescription. National surveillance with a multi-sectoral committee in the "One Health" concept would be a useful measure for monitoring antibiotic use in animals and humans.

2.3 Transmissions dynamics of AMR in developing countries

Due to the unregulated use of antibiotics in agricultural production systems in developing countries, bacteria become resistance to single or multiple antimicrobials. These resistance bacteria or genes are transmitted directly from agricultural food products such as meat, milk, egg, fish, and vegetables to humans. Hospital effluents, garbage, livestock effluents contaminated with resistance bacteria drained to the nearby water body where fishes raised, and this water is also used in the crop fields for their productions. It is another way to transmit resistance bacteria from crops and fish to humans. The fate of AMR bacteria in the agricultural production system and environment is still unclear. Could AMR bacteria and mobile genetic elements carrying the resistance genes further evolve after their transfer to the environment? There are knowledge gaps regarding the magnitude and dynamic nature of spread regarding antimicrobial resistance bacteria and antimicrobial resistance genes within and between different ecological niches on farms, which deserve to be considered when assessing antimicrobial resistance bacteria's transmission the food chain. Moreover, the transmission pathway of resistance bacteria between the agricultural production systems, environment, and humans in developing countries is very complex and given in **Figure 1**.

3. Conclusions

Antimicrobial resistance has shown a profound surge in developing countries as well as around the globe. In developing countries, antibiotic resistance on different microorganisms, especially *E. coli*, *Salmonella* spp. and *Staphylococcus* spp. are skyrocketing in different agricultural production systems, environments, and humans due to the poor management and practices, which is truly terrifying. Further studies are required based on the international standard to evaluate AMR nationwide in every developing country. It is essential to sketch a proper multi-sectoral surveillance plan to research, diagnose and execute necessary steps for combating against multi drugs resistance hitch. There is a need for detailed system biology analysis of resistance development *in-situ*. Metagenomic analysis of bacterial pathogens from diverse sources, including hospitals, veterinary clinics, agricultural production systems including live animal production, marketing, processing, and waste water plants, might underline bacterial pathogens' evolution for integrin-mediated resistance gene transfer in resistance evolution. One Health approach by each government among all stakeholders could promote better exercise and antimicrobial stewardship.

Acknowledgements

I acknowledge the Department of Physiology, Biochemistry and Pharmacology, Faculty of Veterinary Medicine, Chattogram Veterinary and Animal Sciences University, Bangladesh, to contribute my research. I also acknowledge Shahneaz, Mazhar, Shaikat, Mahabub, Tanzin, Nayem, and Kaisar for their help in writing and checking the document. Finally, I acknowledge the Bangladesh Bureau of Educational Information and Statistics (BANBEIS) project number: SD 2019967 for funding.

Conflict of interest

Not exist.

Author details

Mohammad Mahmudul Hassan
Department of Physiology, Biochemistry and Pharmacology, Faculty of Veterinary Medicine, Chattogram Veterinary and Animal Sciences University, Khulshi, Chattogram, Bangladesh

*Address all correspondence to: miladhasan@yahoo.com

References

[1] Shibl A, Memish Z, Osoba A. Antibiotic resistance in developing countries. Journal of chemotherapy. 2001;13(sup1): 40–4.

[2] Appelbaum PC. Microbiology of antibiotic resistance in *Staphylococcus aureus*. Clinical Infectious Diseases. 2007;45(Supplement_3):S165-S70.

[3] Khan SA, Imtiaz MA, Sayeed MA, Shaikat AH, Hassan MM. Antimicrobial resistance pattern in domestic animal-wildlife-environmental niche via the food chain to humans with a Bangladesh perspective; a systematic review. BMC Veterinary Research. 2020;16(1):1–13.

[4] Reinthaler F, Posch J, Feierl G, Wüst G, Haas D, Ruckenbauer G, et al. Antibiotic resistance of *E. coli* in sewage and sludge. Water research. 2003;37(8): 1685–90.

[5] Erb A, Stürmer T, Marre R, Brenner H. Prevalence of antibiotic resistance in Escherichia coli: overview of geographical, temporal, and methodological variations. European Journal of Clinical Microbiology & Infectious Diseases. 2007;26(2):83–90.

[6] Sáenz Y, Zarazaga M, Briñas L, Lantero M, Ruiz-Larrea F, Torres C. Antibiotic resistance in Escherichia coli isolates obtained from animals, foods and humans in Spain. International journal of antimicrobial agents. 2001;18 (4):353–8.

[7] Sarker MS, Mannan MS, Ali MY, Bayzid M, Ahad A, Bupasha ZB. Antibiotic resistance of Escherichia coli isolated from broilers sold at live bird markets in Chattogram, Bangladesh. Journal of advanced veterinary and animal research. 2019;6(3):272.

[8] Akond MA, Alam S, Hassan S, Shirin M. Antibiotic resistance of Escherichia coli isolated from poultry and poultry environment of Bangladesh. Internet Journal of Food Safety. 2009;11: 19–23.

[9] Bouchrif B, Paglietti B, Murgia M, Piana A, Cohen N, Ennaji MM, et al. Prevalence and antibiotic-resistance of Salmonella isolated from food in Morocco. The Journal of Infection in Developing Countries. 2009;3(01): 035–40.

[10] Santos RL, Zhang S, Tsolis RM, Kingsley RA, Adams LG, Bäumler AJ. Animal models of Salmonella infections: enteritis versus typhoid fever. Microbes and infection. 2001;3(14–15):1335–44.

[11] Zhang-Barber L, Turner A, Barrow P. Vaccination for control of Salmonella in poultry. Vaccine. 1999;17 (20–21):2538–45.

[12] Wray C, Sojka W. Bovine salmonellosis. Journal of Dairy Research. 1977;44(2):383–425.

[13] Parvez MAK, Ferdous RN, Rahman MS, Islam S. Healthcare-associated (HA) and community-associated (CA) methicillin resistant *Staphylococcus aureus* (MRSA) in Bangladesh–Source, diagnosis and treatment. Journal of Genetic Engineering and Biotechnology. 2018;16 (2):473–8.

[14] Islam T, Kubra K, Chowdhury MMH. Prevalence of methicillin-resistant *Staphylococcus aureus* in hospitals in Chittagong, Bangladesh: A threat of nosocomial infection. Journal of microscopy and ultrastructure. 2018;6(4):188.

[15] Shrestha LB, Baral R, Poudel P, Khanal B. Clinical, etiological and antimicrobial susceptibility profile of pediatric urinary tract infections in a tertiary care hospital of Nepal. BMC pediatrics. 2019;19(1):36.

[16] Fatima S, Muhammad IN, Khan MN, Jamil S. Phenotypic expression and prevalence of multi drug resistant extended spectrum beta-lactamase producing Escherichia coli and *Klebsiella pneumoniae* in Karachi, Pakistan. Pak J Pharm Sci. 2018;31(4): 1379–84.

[17] Rasool MS, Siddiqui F, Ajaz M, Rasool SA. Prevalence and antibiotic resistance profiles of Gram negative bacilli associated with urinary tract infections (UTIs) in Karachi, Pakistan. Pakistan Journal of Pharmaceutical Sciences. 2019;32(6).

[18] Jamil B, Gawlik D, Syed MA, Shah AA, Abbasi SA, Müller E, et al. Hospital-acquired methicillin-resistant *Staphylococcus aureus* (MRSA) from Pakistan: molecular characterisation by microarray technology. European Journal of Clinical Microbiology & Infectious Diseases. 2018;37(4): 691–700.

[19] Wajid M, Awan AB, Saleemi MK, Weinreich J, Schierack P, Sarwar Y, et al. Multiple drug resistance and virulence profiling of *Salmonella enterica* serovars Typhimurium and Enteritidis from poultry farms of Faisalabad, Pakistan. Microbial Drug Resistance. 2019;25(1):133–42.

[20] Azam M, Mohsin M, Saleemi MK. Virulence-associated genes and antimicrobial resistance among avian pathogenic Escherichia coli from colibacillosis affected broilers in Pakistan. Tropical animal health and production. 2019;51(5):1259–65.

[21] Lv J, Mohsin M, Lei S, Srinivas S, Wiqar RT, Lin J, et al. Discovery of a mcr-1-bearing plasmid in commensal colistin-resistant Escherichia coli from healthy broilers in Faisalabad, Pakistan. Virulence. 2018;9(1):994–9.

[22] Syed MA, Shah SHH, Sherafzal Y, Shafi-ur-Rehman S, Khan MA, Barrett JB, et al. Detection and molecular characterization of methicillin-resistant *Staphylococcus aureus* from table eggs in Haripur, Pakistan. Foodborne pathogens and disease. 2018;15(2):86–93.

[23] Maalik A, Shahzad A, Iftikhar A, Rizwan M, Ahmad H, Khan I. Prevalence and Antibiotic Resistance of *Staphylococcus aureus* and Risk Factors for Bovine Subclinical Mastitis in District Kasur, Punjab, Pakistan. Pakistan Journal of Zoology. 2019;51(3): 1123.

[24] Aqib AI, Nighat S, Rais A, Sana S, Jamal MA, Kulyar MF-e-A, et al. Drug susceptibility profile of *Staphylococcus aureus* isolated from mastitic milk of goats and risk factors associated with goat mastitis in Pakistan. Pakistan Journal of Zoology. 2019;51(1).

[25] Singh AK, Das S, Singh S, Gajamer VR, Pradhan N, Lepcha YD, et al. Prevalence of antibiotic resistance in commensal Escherichia coli among the children in rural hill communities of Northeast India. PloS one. 2018;13(6).

[26] Vigi C, Gaurav S, Naveen C, Raghuvanshi R. High prevalence of multiple drug resistance among pediatric Escherichia Coli infections. International Journal of Medical Research & Health Sciences. 2018;5(10): 166–9.

[27] Mahalingam N, Manivannan B, Khamari B, Siddaramappa S, Adak S, Bulagonda EP. Detection of antibiotic resistance determinants and their transmissibility among clinically isolated carbapenem-resistant Escherichia coli from South India. Medical Principles and Practice. 2018;27 (5):428–35.

[28] Mohsin J, Pál T, Petersen JE, Darwish D, Ghazawi A, Ashraf T, et al. Plasmid-mediated colistin resistance gene mcr-1 in an Escherichia coli ST10

bloodstream isolate in the Sultanate of Oman. Microbial Drug Resistance. 2018; 24(3):278–82.

[29] Purohit MR, Lindahl LF, Diwan V, Marrone G, Lundborg CS. High levels of drug resistance in commensal *E. coli* in a cohort of children from rural central India. Scientific reports. 2019;9(1):1–11.

[30] Shanthi B, Selvi R, Madhumathy A. Antimicrobial susceptibility pattern of Escherichia coli from patients with urinary tract infections in a tertiary care hospital. Int J Curr Microbiol App Sci. 2018;7(01):289–94.

[31] Veeraraghavan B, Walia K. Antimicrobial susceptibility profile & resistance mechanisms of Global Antimicrobial Resistance Surveillance System (GLASS) priority pathogens from India. The Indian Journal of Medical Research. 2019;149(2):87.

[32] Klemm EJ, Shakoor S, Page AJ, Qamar FN, Judge K, Saeed DK, et al. Emergence of an extensively drug-resistant *Salmonella enterica* serovar Typhi clone harboring a promiscuous plasmid encoding resistance to fluoroquinolones and third-generation cephalosporins. MBio. 2018;9(1): e00105–18.

[33] Chatham-Stephens K, Medalla F, Hughes M, Appiah GD, Aubert RD, Caidi H, et al. Emergence of extensively drug-resistant Salmonella Typhi infections among travelers to or from Pakistan—United States, 2016–2018. Morbidity and Mortality Weekly Report. 2019;68(1):11.

[34] Tahir MF, Afzal F, Athar M. Prevalence and Antimicrobial Susceptibility Patterns of Salmonella Enteritidis and Salmonella Typhimurium Isolates from Commercial Poultry in Punjab, Pakistan. Iproceedings. 2018;4(1):e10639.

[35] Engsbro AL, Jespersen HSR, Goldschmidt MI, Mollerup S, Worning P, Pedersen MS, et al. Ceftriaxone-resistant *Salmonella enterica* serotype Typhi in a pregnant traveller returning from Karachi, Pakistan to Denmark, 2019. Eurosurveillance. 2019; 24(21).

[36] Bhave S, Kolhe R, Mahadevaswamy R, Bhong C, Jadhav S, Nalband S, et al. Phylogrouping and antimicrobial resistance analysis of extraintestinal pathogenic Escherichia coli isolated from poultry species. Turkish Journal of Veterinary and Animal Sciences. 2019;43(1):117–26.

[37] Subedi M, Luitel H, Devkota B, Bhattarai RK, Phuyal S, Panthi P, et al. Antibiotic resistance pattern and virulence genes content in avian pathogenic Escherichia coli (APEC) from broiler chickens in Chitwan, Nepal. BMC veterinary research. 2018; 14(1):113.

[38] Majhi M, Pamia J, Panda SK, Samal L, Mishra R. Effect of Age and Season on Enteritis and Antibiotic Sensitivity Test of *E. coli* Isolated from Infected Chickens in Odisha, India. Int J Curr Microbiol App Sci. 2018;7(3): 2037–45.

[39] Magray S, Wani S, Kashoo Z, Bhat M, Adil S, Farooq S, et al. Serological diversity, molecular characterisation and antimicrobial sensitivity of avian pathogenic Escherichia coli (APEC) isolates from broiler chickens in Kashmir, India. Animal production science. 2019;59(2): 338–46.

[40] Zhang J, Chen L, Wang J, Yassin AK, Butaye P, Kelly P, et al. Molecular detection of colistin resistance genes (mcr-1, mcr-2 and mcr-3) in nasal/oropharyngeal and anal/cloacal swabs from pigs and poultry. Scientific reports. 2018;8(1):1–9.

[41] Jamoh K, Rajkhowa T, Singh Y, Ravindran R, Arya R. Antimicrobial

resistant Escherichia coli and associated colibacillosis in poultry population of Mizoram. Indian Journal of Veterinary Pathology. 2018;42(3):185–90.

[42] Khoirani K, Indrawati A, Setiyaningsih S. Detection of Ampicillin Resistance Encoding Gene of Escherichia coli from Chickens in Bandung and Purwakarta. Jurnal Riset Veteriner Indonesia (Journal of The Indonesian Veterinary Research). 2019;3(1).

[43] Enany M, Hassan W, Ismail N. Prevalence of Antibiotic Resistance Genes among *E. coli* Strains Isolated from Poultry in Suez Canal Area. Suez Canal Veterinary Medicine Journal SCVMJ. 2018;23(1):53–65.

[44] Maru V, Ranade V. Antibiotic resistant-biofilm forming Escherichia coli and Salmonella spp. in poultry raw meat. Indian Journal of Veterinary Research (The). 2018;27(2):33–8.

[45] Waghamare R, Paturkar A, Vaidya V, Zende R, Dubal Z, Dwivedi A, et al., editors. Phenotypic and genotypic drug resistance profile of Salmonella serovars isolated from poultry farm and processing units located in and around Mumbai city, India, Veterinary World, 11 (12): 1682–16882018: Abstract.

[46] Zehra A, Gulzar M, Singh R, Kaur S, Gill J. Prevalence, multidrug resistance and molecular typing of methicillin-resistant *Staphylococcus aureus* (MRSA) in retail meat from Punjab, India. Journal of global antimicrobial resistance. 2019;16:152–8.

[47] Ruban SW, Babu RN, Abraham RJ, Senthilkumar T, Kumaraswamy P, Porteen K, et al. Prevalence and Antimicrobial Susceptibility of *Staphylococcus aureus* Isolated from Retail Chicken Meat in Chennai, India. Journal of Animal Research. 2018;8(3): 423–7.

[48] Murugan MS, Sinha D, Kumar OV, Yadav AK, Pruthvishree B, Vadhana P, et al. Epidemiology of carbapenem-resistant Escherichia coli and first report of blaVIM carbapenemases gene in calves from India. Epidemiology & Infection. 2019;147.

[49] Shah MS, Qureshi S, Kashoo Z, Farooq S, Wani SA, Hussain MI, et al. Methicillin resistance genes and in vitro biofilm formation among *Staphylococcus aureus* isolates from bovine mastitis in India. Comparative immunology, microbiology and infectious diseases. 2019;64:117–24.

[50] Yadav R, Kumar A, Singh VK, Yadav SK. Prevalence and antibiotyping of *Staphylococcus aureus* and methicillin-resistant *S. aureus* (MRSA) in domestic animals in India. Journal of global antimicrobial resistance. 2018;15:222–5.

[51] Ruban SW, Babu RN, Abraham RJ, Senthilkumar T, Kumraswamy P, Rao VA. Prevalence of methicillin resistant *Staphylococcus aureus* in retail buffalo meat in Chennai, India. Buffalo Bulletin. 2018;37(1):51–8.

[52] Dutta TK, Chakraborty S, Das M, Mandakini R. Multidrug-resistant Staphylococcus pettenkoferi isolated from cat in India. Veterinary world. 2018;11(10):1380.

[53] Ghafur A, Shankar C, GnanaSoundari P, Venkatesan M, Mani D, Thirunarayanan M, et al. Detection of chromosomal and plasmid-mediated mechanisms of colistin resistance in Escherichia coli and *Klebsiella pneumoniae* from Indian food samples. Journal of global antimicrobial resistance. 2019;16:48–52.

[54] Batabyal K, Banerjee A, Pal S, Dey S, Joardar SN, Samanta I, et al. Detection, characterization, and antibiogram of extended-spectrum beta-lactamase Escherichia coli isolated from bovine milk samples in West Bengal, India. Veterinary world. 2018;11(10):1423.

[55] Patel R, Kumar R, Savalia C, Patel N. Isolation of *Staphylococcus aureus* from Raw Cattle Milk and their Drug Resistance Pattern. Int J Curr Microbiol App Sci. 2018;7(2):836–40.

[56] Sivakumar M, Dubal ZB, Kumar A, Bhilegaonkar K, Kumar ORV, Kumar S, et al. Virulent methicillin resistant *Staphylococcus aureus* (MRSA) in street vended foods. Journal of food science and technology. 2019;56(3):1116–26.

[57] Sharma P. Water Quality of River Narmada at Gwari Ghat Jabalpur (MP, India) in Terms of Microbial Load, Drug Resistance and Potability. Journal of Applied & Environmental Microbiology. 2018;6(1):25–9.

[58] Dhawde R, Macaden R, Saranath D, Nilgiriwala K, Ghadge A, Birdi T. Antibiotic resistance characterization of environmental *E. coli* isolated from River Mula-Mutha, Pune District, India. International journal of environmental research and public health. 2018; 15(6):1247.

[59] Odonkor ST, Addo KK. Prevalence of multidrug-resistant Escherichia coli isolated from drinking water sources. International journal of microbiology. 2018;2018.

[60] Rayasam SD, Ray I, Smith KR, Riley LW. Extraintestinal pathogenic escherichia coli and antimicrobial drug resistance in a maharashtrian drinking water system. The American journal of tropical medicine and hygiene. 2019;100 (5):1101–4.

[61] Asaduzzaman M, Baral K, Islam MM, Nayem A, Alam J, Juliana FM, et al. Susceptibility pattern of second line antibiotic colistin against gram negative bacteria causing urinary tract infection in selected areas Dhaka city, Bangladesh. Eur J Biomed Pharm Sci. 2018;5(3):874–9.

[62] Asaduzzaman M, Miah AA, Bhuiyan M, Alam J, Juliana F, Hossain N. Resistant pattern of nalidixic acid against uropathogens in selected areas of Dhaka city, Bangladesh. Eur J Biomed Pharm Sci. 2018;5(3):90–5.

[63] Asaduzzaman M, Asaduzzaman Shamim M, Mian S, Alam MJ, Juliana FM, Hossain N, et al. Resistance pattern of cefixime against uropathogens causing urinary tract infection in selected areas of Dhaka city, Bangladesh. Int J Eng Sci. 2018;7 (1):33–9.

[64] Acherjya GK, Tarafder K, Ghose R, Islam DU, Ali M, Akhtar N, et al. Pattern of Antimicrobial Resistance to Escherichia Coli Among the Urinary Tract Infection Patients in Bangladesh. American Journal of Internal Medicine. 2018;6(5):132–7.

[65] Asaduzzaman M, Hasan MZ, Khatun M, Alam J, Hossain N, Das B, et al. Resistance pattern of levofloxacin against uropathogens causing urinary tract infection in selected areas of Dhaka city. Bangladesh J Biol Agri Healthc. 2018;8(4):74–81.

[66] Asaduzzaman M, Ullah MM, Redwan S, Alam J, Juliana FM, Hossain N, et al. Emergence of meropenem resistance in pathogens recovered from urine cultures in Bangladesh. IOSR JPBS. 2018;13(3):41–7.

[67] Tanmoy AM, Westeel E, De Bruyne K, Goris J, Rajoharison A, Sajib MS, et al. *Salmonella enterica* Serovar Typhi in Bangladesh: exploration of genomic diversity and antimicrobial resistance. MBio. 2018;9 (6):e02112–18.

[68] Ahsan S, Rahman S. Azithromycin resistance in clinical isolates of *Salmonella enterica* serovars Typhi and paratyphi in Bangladesh. Microbial Drug Resistance. 2019;25(1):8–13.

[69] Jahan M, Rahman M, Rahman M, Sikder T, Uson-Lopez RA, Selim ASM,

et al. Microbiological safety of street-vended foods in Bangladesh. Journal of Consumer Protection and Food Safety. 2018;13(3):257–69.

[70] Banik A, Abony M, Datta S, Towhid ST. Microbiological quality of ready-to-eat food from Dhaka, Bangladesh. Current Research in Nutrition and Food Science Journal. 2019;7(1):161–8.

[71] Rahman M, Rahman A, Islam M, Alam M. Detection of multi–drug resistant Salmonella from milk and meat in Bangladesh. Bangladesh Journal of Veterinary Medicine. 2018;16(1):115–20.

[72] Hasan M, Kabir SL, Rahman T, Sarker YA. Bacteriological quality assessment of buffalo meat collected from different districts of Bangladesh with particular emphasis on the molecular detection and antimicrobial resistance of the isolated Salmonella species. Asian Australas. J Food Saf Secur. 2018;2:12–20.

[73] Rahman MA, Rahman AA, Islam MA, Alam MM. Multi–drug resistant *Staphylococcus aureus* isolated from milk, chicken meat, beef and egg in Bangladesh. Research in Agriculture Livestock and Fisheries. 2018;5(2):175–83.

[74] Hoque M, Das Z, Rahman A, Haider M, Islam M. Molecular characterization of *Staphylococcus aureus* strains in bovine mastitis milk in Bangladesh. International journal of veterinary science and medicine. 2018;6 (1):53–60.

[75] Debnath T, Bhowmik S, Islam T, Chowdhury MMH. Presence of multidrug-resistant bacteria on mobile phones of healthcare workers accelerates the spread of nosocomial infection and Regarded as a Threat to Public Health in Bangladesh. Journal of microscopy and ultrastructure. 2018; 6(3):165.

[76] Hassan MS, Kabir SL, Sarker YA, Rahman MT. Bacteriological assessment of tap water collected from different markets of Mymensingh, Gazipur and Sherpur districts of Bangladesh with special focus on the molecular detection and antimicrobial resistance of the isolated Escherichia coli. Asian Australas. J Food Saf Secur. 2018;2(1): 21–8.

[77] Anwar T. Determination of prevalence and antibiotic susceptibility pattern of bacteria isolated from household and restaurant kitchen utensils of Dhaka, Bangladesh: BRAC Univeristy; 2018.

[78] Whistler T, Sapchookul P, McCormick DW, Sangwichian O, Jorakate P, Makprasert S, et al. Epidemiology and antimicrobial resistance of invasive non-typhoidal Salmonellosis in rural Thailand from 2006–2014. PLoS neglected tropical diseases. 2018;12(8):e0006718.

[79] Luk-In S, Chatsuwan T, Pulsrikarn C, Bangtrakulnonth A, Rirerm U, Kulwichit W. High prevalence of ceftriaxone resistance among invasive *Salmonella enterica* serotype Choleraesuis isolates in Thailand: the emergence and increase of CTX-M-55 in ciprofloxacin-resistant S. Choleraesuis isolates. International Journal of Medical Microbiology. 2018; 308(4):447–53.

[80] Jaganath D, Jorakate P, Makprasert S, Sangwichian O, Akarachotpong T, Thamthitiwat S, et al. *Staphylococcus aureus* bacteremia incidence and methicillin resistance in rural Thailand, 2006–2014. The American journal of tropical medicine and hygiene. 2018;99(1):155–63.

[81] Kitti T, Seng R, Saiprom N, Thummeepak R, Chantratita N, Boonlao C, et al. Molecular characteristics of methicillin-resistant staphylococci clinical isolates from a

tertiary Hospital in Northern Thailand. Canadian Journal of Infectious Diseases and Medical Microbiology. 2018;2018.

[82] Pootong A, Mungkornkeaw N, Norrapong B, Cowawintaweewat S. Phylogenetic background, drug susceptibility and virulence factors of uropathogenic *E. coli* isolate in a tertiary university hospital in central Thailand. Tropical Biomedicine. 2018;35(1): 195–204.

[83] Eiamphungporn W, Yainoy S, Jumderm C, Tan-arsuwongkul R, Tiengrim S, Thamlikitkul V. Prevalence of the colistin resistance gene mcr-1 in colistin-resistant Escherichia coli and *Klebsiella pneumoniae* isolated from humans in Thailand. Journal of global antimicrobial resistance. 2018;15:32–5.

[84] Prasertsiriphong S, Chootong R, Jamulitrat S, Penghmak M. Prevalence of Antibiotic Resistance in Escherichia coli from the Fecal Flora of Humans in a Rural Area of Songkhla Province. Journal of Health Science and Medical Research. 2019:321–7.

[85] Dahal RH, Chaudhary DK. Microbial infections and antimicrobial resistance in Nepal: current trends and recommendations. The open microbiology journal. 2018;12:230.

[86] Pokhrel B, Koirala T, Shah G, Joshi S, Baral P. Bacteriological profile and antibiotic susceptibility of neonatal sepsis in neonatal intensive care unit of a tertiary hospital in Nepal. BMC pediatrics. 2018;18(1):208.

[87] Yadav NS, Sharma S, Chaudhary DK, Panthi P, Pokhrel P, Shrestha A, et al. Bacteriological profile of neonatal sepsis and antibiotic susceptibility pattern of isolates admitted at Kanti Children's Hospital, Kathmandu, Nepal. BMC research notes. 2018;11(1):301.

[88] Pumipuntu N, Tunyong W, Chantratita N, Diraphat P, Pumirat P, Sookrung N, et al. Staphylococcus spp. associated with subclinical bovine mastitis in central and northeast provinces of Thailand. PeerJ. 2019;7: e6587.

[89] Tansawai U, Sanguansermsri D, Na-udom A, Walsh TR, Niumsup PR. Occurrence of extended spectrum β-lactamase and AmpC genes among multidrug-resistant Escherichia coli and emergence of ST131 from poultry meat in Thailand. Food control. 2018;84: 159–64.

[90] Nuangmek A, Rojanasthien S, Chotinun S, Yamsakul P, Tadee P, Thamlikitkul V, et al. Antimicrobial resistance in ESBL-producing Escherichia coli isolated from layer and pig farms in Thailand. Acta Scientiae Veterinariae. 2018;46(1):8.

[91] Sripaurya B, Ngasaman R, Benjakul S, Vongkamjan K. Virulence genes and antibiotic resistance of Salmonella recovered from a wet market in Thailand. Journal of Food Safety. 2019;39(2):e12601.

[92] Pongsilp N, Nimnoi P. Diversity and antibiotic resistance patterns of enterobacteria isolated from seafood in Thailand. CyTA-Journal of Food. 2018; 16(1):793–800.

[93] Intrakamhaeng M, Singpun Y, Sreward C, Phakhunthod S, Ketphonthong S. The occurrence of MRSA, MSSA and antibiotic resistance, related factors in area of dairy farming of Mahasarakham province, Thailand. International Journal of Agricultural Technology. 2018;14(7 Special Issue): 1259–66.

[94] Fukuda A, Usui M, Okubo T, Tagaki C, Sukpanyatham N, Tamura Y. Co-harboring of cephalosporin (bla)/ colistin (mcr) resistance genes among Enterobacteriaceae from flies in Thailand. FEMS microbiology letters. 2018;365(16):fny178.

[95] Thapa S, Sapkota LB. Changing trend of neonatal septicemia and antibiotic susceptibility pattern of isolates in Nepal. International journal of pediatrics. 2019;2019.

[96] Manandhar S, Singh A, Varma A, Pandey S, Shrivastava N. Biofilm producing clinical *Staphylococcus aureus* isolates augmented prevalence of antibiotic resistant cases in tertiary care hospitals of Nepal. Frontiers in microbiology. 2018;9:2749.

[97] Mahato S, Mahato A, Yadav J. Prevalence and identification of uropathogens in eastern Nepal and understanding their antibiogram due to multidrug resistance and Esbl. Asian Pac J Microbiol Res. 2018;2(1):09–17.

[98] Britto CD, Dyson ZA, Duchene S, Carter MJ, Gurung M, Kelly DF, et al. Laboratory and molecular surveillance of paediatric typhoidal Salmonella in Nepal: Antimicrobial resistance and implications for vaccine policy. PLoS neglected tropical diseases. 2018;12(4): e0006408.

[99] Margulieux KR, Srijan A, Ruekit S, Nobthai P, Poramathikul K, Pandey P, et al. Extended-spectrum β-lactamase prevalence and virulence factor characterization of enterotoxigenic Escherichia coli responsible for acute diarrhea in Nepal from 2001 to 2016. Antimicrobial Resistance & Infection Control. 2018;7(1):87.

[100] Neopane P, Nepal HP, Shrestha R, Uehara O, Abiko Y. In vitro biofilm formation by *Staphylococcus aureus* isolated from wounds of hospital-admitted patients and their association with antimicrobial resistance. International journal of general medicine. 2018;11:25.

[101] Khanal LK, Adhikari RP, Guragain A. Prevalence of Methicillin Resistant *Staphylococcus aureus* and Antibiotic Susceptibility Pattern in a Tertiary Hospital in Nepal. Journal of Nepal Health Research Council. 2018;16 (2):172–4.

[102] Petersiel N, Shresta S, Tamrakar R, Koju R, Madhup S, Shresta A, et al. The epidemiology of typhoid fever in the Dhulikhel area, Nepal: A prospective cohort study. PloS one. 2018;13(9).

[103] Shrestha R, Khanal S, Poudel P, Khadayat K, Ghaju S, Bhandari A, et al. Extended spectrum β-lactamase producing uropathogenic Escherichia coli and the correlation of biofilm with antibiotics resistance in Nepal. Annals of Clinical Microbiology and Antimicrobials. 2019;18(1):42.

[104] Wagle S, Khanal BR, Tiwari BR. High susceptibility of fosfomycin to uropathogenic Escherichia coli isolated at Tertiary Care Hospital of Nepal. Journal of Advances in Microbiology. 2018:1–8.

[105] Roberts MC, Joshi PR, Greninger AL, Melendez D, Paudel S, Acharya M, et al. The human clone ST22 SCC mec IV methicillin-resistant *Staphylococcus aureus* isolated from swine herds and wild primates in Nepal: is man the common source? FEMS microbiology ecology. 2018;94(5): fiy052.

[106] Maharjan A, Bhetwal A, Shakya S, Satyal D, Shah S, Joshi G, et al. Ugly bugs in healthy guts! Carriage of multidrug-resistant and ESBL-producing commensal Enterobacteriaceae in the intestine of healthy Nepalese adults. Infection and drug resistance. 2018;11:547.

[107] Gurung RR, Maharjan P, Chhetri GG. Antibiotic resistance pattern of *Staphylococcus aureus* with reference to MRSA isolates from pediatric patients. Future Science OA. 2020(0):FSO464.

[108] Subedi M, Bhattarai RK, Devkota B, Phuyal S, Luitel H.

Correction to: Antibiotic resistance pattern and virulence genes content in avian pathogenic Escherichia coli (APEC) from broiler chickens in Chitwan, Nepal. BMC veterinary research. 2018;14(1):166.

[109] Bantawa K, Sah SN, Limbu DS, Subba P, Ghimire A. Antibiotic resistance patterns of *Staphylococcus aureus*, Escherichia coli, Salmonella, Shigella and Vibrio isolated from chicken, pork, buffalo and goat meat in eastern Nepal. BMC research notes. 2019;12(1):1–6.

[110] Saud B, Paudel G, Khichaju S, Bajracharya D, Dhungana G, Awasthi MS, et al. Multidrug-resistant bacteria from raw meat of buffalo and chicken, Nepal. Veterinary medicine international. 2019;2019.

[111] Dabo NT, Muhammad B, Saka HK, Kalgo ZM, Raheem RA. Antibiotic Resistance Pattern of Escherichia coli Isolated from Diarrhoeic and Non-diarrhoeic Under Five Children in Kano, Nigeria. Journal of Microbiology and Biotechnology. 2019;4(3):94–102.

[112] Osungunna MO, Onawunmi GO. Antibiotic resistance profiles of biofilm-forming bacteria associated with urine and urinary catheters in a tertiary hospital in Ile-Ife, Nigeria. Southern African Journal of Infectious Diseases. 2018;33(3):80–5.

[113] Omoyibo EE, Oladele AO, Ibrahim MH, Adekunle OT. Antibiotic susceptibility of wound swab isolates in a tertiary hospital in Southwest Nigeria. Annals of African medicine. 2018; 17(3):110.

[114] Oli AN, Ogbuagu VI, Ejikeugwu CP, Iroha IR, Ugwu MC, Ofomata CM, et al. Multi-Antibiotic Resistance and Factors Affecting Carriage of Extended Spectrum β-Lactamase-Producing Enterobacteriaceae in Pediatric Population of Enugu Metropolis, Nigeria. Medical Sciences. 2019;7(11): 104.

[115] Mustapha A, Imir T. Detection of Multidrug-Resistance Gram-Negative Bacteria from Hospital Sewage in North East, Nigeria. Frontiers in Environmental Microbiology. 2019;5(1):1.

[116] Akinyemi KO, Oyefolu AOB, Mutiu WB, Iwalokun BA, Ayeni ES, Ajose SO, et al. Typhoid fever: tracking the trend in Nigeria. The American journal of tropical medicine and hygiene. 2018;99(3_Suppl):41–7.

[117] Eghieye M, Jodi S, Bassey B, Nkene I, Abimiku R, Ngwai Y. Antimicrobial resistance profile of Escherichia coli isolated from urine of patients in selected General Hospitals in Abuja Municipal, Nigeria. Asian Journal of Advanced Research and Reports. 2018:1–10.

[118] Osiyemi J, Osinupebi O, Ejilude O, Makanjuola S, Sunmola N, Osiyemi E. Antibiotic Resistance Profile of Methicillin-Resistant *Staphylococcus aureus* in Abeokuta, Nigeria. Journal of Advances in Microbiology. 2018:1–9.

[119] Onanuga A, Omeje MC, Eboh DD. Carriage of multi-drug resistant urobacteria by asymptomatic pregnant women in Yenagoa, Bayelsa State, Nigeria. African journal of infectious diseases. 2018;12(2):14–20.

[120] Alechenu EC, Nweze JA, Lerum NI, Eze EA. Prevalence and Antibiotic Resistance Patterns of Gram-Negative Uropathogens among Paediatric Patients in Nigeria. Open Journal of Medical Microbiology. 2019;9 (04):215.

[121] Adelowo OO, Caucci S, Banjo OA, Nnanna OC, Awotipe EO, Peters FB, et al. Extended Spectrum Beta-Lactamase (ESBL)-producing bacteria isolated from hospital wastewaters,

rivers and aquaculture sources in Nigeria. Environmental Science and Pollution Research. 2018;25(3):2744–55.

[122] Awogbemi J, Adeyeye M, Akinkunmi E. A Survey of Antimicrobial Agents Usage in Poultry Farms and Antibiotic Resistance in Escherichia Coli and Staphylococci Isolates from the Poultry in Ile-Ife. Journal of Infectious Diseases and Epidemiology. 2018;4(1).

[123] Kwoji I, Tambuwal F, Abubakar M, Yakubu Y, Musa J, Jauro S, et al. Antibiotic sensitivity patterns of methicillin-resistant staph-ylococcus aureus isolated from chickens in poultry farms in sokoto, nigeria. Adv Anim Vet Sci. 2018;6(1):8–11.

[124] Oloso NO, Fagbo S, Garbati M, Olonitola SO, Awosanya EJ, Aworh MK, et al. Antimicrobial resistance in food animals and the environment in Nigeria: A review. International journal of environmental research and public health. 2018;15(6):1284.

[125] Adamu MS, Ugochukwu ICI, Idoko SI, Kwabugge YA, Abubakar NSa, Ameh JA. Virulent gene profile and antibiotic susceptibility pattern of Shiga toxin-producing Escherichia coli (STEC) from cattle and camels in Maiduguri, North-Eastern Nigeria. Tropical animal health and production. 2018;50(6):1327–41.

[126] Onifade A, Afolami O. Antibiotic resistance patterns of Salmonella spp from clinical and water samples in Akure, Ondo State, Nigeria. Asian Journal of Research in Medical and Pharmaceutical Sciences. 2018:1–10.

[127] Abu G, Wondikom A. Isolation, characterization and antibiotic resistance profile studies of bacteria from an excavated pond in Port Harcourt Metropolis, Nigeria. Journal of Applied Sciences and Environmental Management. 2018;22(8):1177–84.

[128] Akaniro I, Oguh C, Kafilat K, Ahmed I, Ezeh C. Physicochemical properties, bacteriological quality and antimicrobial resistance profile of isolates from groundwater sources in ile-ife suburbs, Southwest Nigeria. IOSR Journal of Environmental Science, Toxicology and Food Technology. 2019; 13(1):58–65.

[129] Fowoyo P, Abuo G. Bacteriological Profiling and Antibiotic Resistance of Bacteria Isolated From River Niger Lokoja Tributary, Nigeria. Journal of Applied Life Sciences International. 2018:1–11.

[130] Akinyemi K, Ajoseh S, Iwalokun B, Oyefolu A, Fakorede C, Abegunrin R, et al. Antimicrobial resistance and plasmid profiles of *Salmonella enterica* serovars from different sources in Lagos, Nigeria. Health. 2018;10(6):758–72.

[131] Ajoke AO, Adetokunboh OA. Multiple-Antibiotic Resistance Pattern of Coliform Bacteria Isolated from Different Sources in Iwo, Nigeria. Open Science Journal of Bioscience and Bioengineering. 2018;5(3):41.

[132] Nyandjou Y, Yakubu S, Abdullahi I, Machido D. Multidrug resistance patterns and multiple antibiotic resistance index of salmonella species isolated from Waste Dumps in Zaria Metropolis, Nigeria. Journal of Applied Sciences and Environmental Management. 2019;23(1):41–6.

[133] Borges KA, Furian TQ, Souza SNd, Salle CTP, Moraes HLdS, Nascimento VPd. Antimicrobial resistance and molecular characterization of *Salmonella enterica* serotypes isolated from poultry sources in Brazil. Brazilian Journal of Poultry Science. 2019;21(1).

[134] Cunha-Neto Ad, Carvalho LA, Carvalho RCT, dos Prazeres Rodrigues D, Mano SB, Figueiredo EEdS, et al. Salmonella isolated from chicken carcasses from a slaughterhouse

in the state of Mato Grosso, Brazil: antibiotic resistance profile, serotyping, and characterization by repetitive sequence-based PCR system. Poultry science. 2018;97(4):1373–81.

[135] Moura Q, Fernandes MR, Silva KC, Monte DF, Esposito F, Dropa M, et al. Virulent nontyphoidal Salmonella producing CTX-M and CMY-2 β-lactamases from livestock, food and human infection, Brazil. Virulence. 2018;9(1):281–6.

[136] Almeida F, Seribelli AA, Medeiros MIC, dos Prazeres Rodrigues D, De MelloVarani A, Luo Y, et al. Phylogenetic and antimicrobial resistance gene analysis of Salmonella Typhimurium strains isolated in Brazil by whole genome sequencing. PloS one. 2018;13(8).

[137] da Cunha-Neto A, Carvalho LA, Castro VS, Barcelos FG, Carvalho RCT, Rodrigues DdP, et al. Salmonella Anatum, S. Infantis and S. Schwarzengrund in Brazilian Cheeses: occurrence and antibiotic resistance profiles. International Journal of Dairy Technology. 2020;73(1):296–300.

[138] Acharya KP, Wilson RT. Antimicrobial resistance in Nepal: a review. Frontiers in medicine. 2019;6:105.

[139] Shrivastava SR, Shrivastava PS, Ramasamy J. World health organization releases global priority list of antibiotic-resistant bacteria to guide research, discovery, and development of new antibiotics. Journal of Medical Society. 2018;32(1):76.

[140] Bartley PS, Domitrovic TN, Moretto VT, Santos CS, Ponce-Terashima R, Reis MG, et al. Antibiotic resistance in Enterobacteriaceae from surface waters in urban Brazil highlights the risks of poor sanitation. The American journal of tropical medicine and hygiene. 2019;100(6):1369–77.